Encounters of a World-Traveling Scientist

Tasora

Encounters of a World-Traveling Scientist

Stanley Randolf, M.Sc.

Tasora

Tasora Books
5120 Cedar Lake Road
Minneapolis, MN 55416
(952) 345-4488
Distributed by Itasca Books

Printed in the U.S.A.

Cover design by Debbie Johnson and John Houlgate
Original cover photo by Vinicius-Malta, courtesy Pexels.com
Section-separator design by Irina Zharkova, modified by Debbie Johnson

Publishers Note: These stories are either true chronologically as experienced by the author or expressed as a combination of related events in a different order from what the author experienced.

Disclaimer: Any connection to any person living or deceased is purely coincidental; the names in these stories have been changed.

Library of Congress Cataloging-in-Publication Data

Randolf, Stanley.
Encounters of a World-Traveling Scientist/Stanley Randolf
Summary: Fourteen short stories of mysterious and exciting encounters spanning the globe. Includes back stories from veteran scientists and military throughout the world about strange phenomena. Some sporting adventure included.

ISBN- 978-1-948192-01-9 (softcover)
1. Travel - Non-Fiction 2. Adventure – Non-Fiction 3. Sportsmen Adventure – Non-Fiction

Printed in the United States of America

Dear Reader,

Thank you for taking time to read these short stories; the fourteen stories are based on true experience. This is the third book in a series that follows Adventures *of a World Traveling Scientist* and its companion, *More Adventures of a World Traveling Scientist*. I was present in each one of the forty-five stories, either taking part in the action or interviewing local people that had. When I look back it seems unreal to have encountered such a wide array of experiences; about half were scheduled and half came out of the blue.

Many of the stories detail my love of fishing and safaris, which buffer the deeper, more serious stories that challenge the hippocampus most. I hope you non-anglers read these fishing stories; they touch on more than just catching big fish. After reading one of the more serious stories, like The Helsinki Connection, Under the Rising Sun, Eleven Miles from Sydney, and The Deep State Divide, you may require a break. The stories can be read in any order, so you may want to consider reading a lighter, fishing story, such as Acapulco, or Confrontation in Tahiti, with the heavier stories. My favorite is Tarpon Times in Key West, where I rescue an old fisherman from himself.

I've been fortunate to have travelled the world over four decades, fifty-seven countries last count. My various businesses and business relationships have centered on using beneficial bacteria—probiotics—to provide natural solutions for environmental pollution, crop fertilization, disease control, non-polluting aquaculture, and both animal and human intestinal health. The science provided should not be intellectually overwhelming, I'm confident you'll all handle it.

I owe a world of thanks and gratitude to those people, friends, business associates, and many locals for their contributions. For reasons due to either sensitivity of the information or my desire to not commercialize or offend anyone, only pseudonyms are used.

If I could do it all over, I can't think of anything I'd do differently. When in Rome, do as the Romans, and if you want the adventure of a lifetime, pay attention to your *Soul Voice* for valuable guidance. Mine was loudly commanding me when I was face to face with a monster tiger-shark in Tahiti. Travel with me now and enjoy...

Sincerely,

Stanley Randolf

Stanley Randolf

Table of Contents

Helsinki Harbor and Cathedral

Chapter One

The Helsinki Connection
Antarctic Secrets

Prologue

Vostok Station, Boctok in Russian, was established by the Soviet Union in 1957. The most remote of the Antarctic research stations, it's located in the center of the East Antarctic Ice Sheet (Princess Elizabeth Land), 1,300 kilometers from the South Pole. It's been billed as the coldest spot on Earth, where the temperature reached a record low of -89° C (-128° F) one cheery July day in 1983.

Summer temperatures average -32° C (-25° F). Vostok Station also has an elevation issue that exacerbates its subzero conditions—it's 3,488 meters (11,440 feet) above sea level—making it exceptionally dry and windy.

Twenty-five scientists and engineers, some with their immediate families, typically work at the station in summer, dropping to around fifteen in winter. It's not easy to live this far inland in Antarctica; an acclimation period of a month or more is usually required. From late April to mid-August (winter), eighty-five continuous days of darkness prevail where it's too dark outside to read; the sun remains at least six degrees below the horizon—polar nights dominate.

Due to the extreme stress of living under such conditions, multiple health issues are common: insomnia, headaches, ear pain, nosebleeds, muscle and joint pain, breathing issues, and unworldly hallucinations, to mention a few.

Four thousand meters (2.5 miles) below the ice at Vostok Station, is a 12,500-square-kilometer subglacial lake: Lake Vostok. Close in size to Lake Ontario, it's not frozen and has an average depth of 432 meters. Glaciologists estimate Lake Vostok's fresh water—all 5,400 cubic kilometers of it—has been sealed off under the ice for thousands, possibly millions of years.

The lake has been drilled into by the Russians, and strange discoveries have been reported. Two major surprises were the water temperature of the lake, 15.5° C (60°F) and the presence of previously unknown microbes. A third surprise was the discovery of a thousand-foot open dome above the lake, resulting from water vapor melting a portion of the ice ceiling. Truly an alien realm under the ice.

It was common for the scientists and workers to experience significant unease and bizarre sightings when drilling and doing strenuous work at Vostok Station—arguably the most hostile conditions on Earth.

The Soviet government was aware of many reported instances of strange sightings in Vostok's vicinity: disc and egg-shaped aerial craft that came out of large ice caves, flew around at great speeds, then disappeared over the Transantarctic Mountains. In May of 1987 the Soviet government ordered an investigative team of special operations commandos to Vostok Station. Sergei Sokolov and Viktor Petrov were two of the six commandos assigned to the team, which also included a general and two KGB officers.

Antarctica—February 1987

Sergei Sokolov recorded the following event in a personal notebook the day after it occurred:

"Viktor and I followed General Abramov and the KGB officer on the morning of February 9, 1987. We were on two separate snow machines traveling west of Vostok Station. Our instructions were simply to follow, and when they stopped, we were to stop twenty meters behind them. About thirty kilometers from Vostok we encountered a white haze of fine snow, we had to slow down in the reduced visibility. Shortly, we could see an oval shape on the snow less than a kilometer away.

"The craft was somewhat smaller than a military tank and shaped like a slightly compressed egg. It was dull silver with a smooth metallic surface, had no wings, and was supported in the snow by four tubular legs.

"I didn't see any windows. A panel with a ramp opened, and two skinny beings came out, both about a meter and a half tall. They resembled young oriental boys from our distance. Their faces were human-like but with larger round eyes—dark brown eyes. They appeared to be wearing tight-fitting suits—grey, very thin, like leotards. I wondered if they could feel the cold. We were bundled in heavy, white camouflage parkas; the temperature was minus forty degrees Celsius. The beings appeared to be wearing socks—no shoes or boots. They didn't leave any impressions in the snow.

"General Abramov and the KGB officer drove toward them, stopping about three meters away. Viktor and I were twenty meters back and had our snow machine idling. One of the beings pointed to us then looked at the general, who immediately gave us a sign to shut our machine off. It was clear the beings wanted it shut off.

"The general then opened both his hands, fingers wide apart, palms facing the beings, and made a triangle with his two index fingers and thumbs. The beings responded by tapping a circular patch on their chests which had a triangle in it. The general then said something to them we couldn't hear. After that, nobody talked for twenty minutes. They just stared at each other with the general often nodding his head, like they were communicating telepathically. The beings didn't smile or frown; their facial expressions were emotionless and never changed. I couldn't see the general's face, but the KGB officer appeared startled. Finally, one of the beings—they both looked identical—turned and went back into the craft and brought out a metallic yellow tablet about the dimensions of a standard envelope. The being held it up for the general to see and turned it on. It lit up like an LED screen on a computer, but in brilliant colors which projected three-dimensional images from the screen. The being took only ten minutes to show the general how to operate it, then handed it to him, stared for a moment at the KGB officer, looked briefly at Viktor and me, then returned with the other being to their craft.

"The general and KGB officer walked back to their machine. Viktor was about to start our snow machine when the general barked '*Het!*' (No) and pointed up. He didn't want the machines running until the craft had departed. Shortly, the egg-shaped craft rose slowly to about fifty meters, then accelerated vertically and disappeared without a sound. The hazy conditions cleared up.

"The general felt his parka's zippered pocket, presumably confirming the yellow tablet's location. It was obvious this was not his first meeting with these beings. The KGB agent, who never said one word, sat down beside the general. We sped back to Vostok Station.

"The next morning the KGB officer that had been with us was not at breakfast. The general met me in the lavatory and told me the officer had committed suicide that night; he was found dead in one of the tunnels that connected the buildings. He had shot himself in the head. I thought this was very strange; KGB officers were trained to be imperturbable. Maybe he hadn't acclimated to the stress of Antarctica? General Abramov looked deep into my eyes, paused, then ordered me to never mention the meeting with the two beings. He gave Viktor the same order. Officially, the meeting 'never happened.' He told me to plan to meet with him next month in Moscow—that I might be required for a critical assignment. I could feel his stress and see the effects the Cold War was having on him. More than a dozen countries were spying on us, including our government." *End of Sergei Sokolov's notebook entry.*

Northwestern Russia, February 1990

Sergei Sokolov and Viktor Petrov telemarked downhill wearing white camouflage jumpsuits with hoods and goggles pulled tight—Russian Spetsnaz thermal wear for arctic duty. They were well practiced at military-style cross country skiing. It was a brisk winter night on the Kola Peninsula in northwestern Russia. Each of the retired commandos was armed with a 9mm-7N21 pistol that they were prepared to use on fellow comrades if necessary. When the downhill slope quit and the forested terrain flattened, they double-poled to a stand of snow-draped spruce trees and stopped. Iridescent curtains of the aurora borealis illuminated the night.

"Big improvement over a night at Vostok Station," Sergei quipped.

"This is a heatwave," Viktor replied. "At Vostok, our vodka froze!" The average temperature in February (summer) at Vostok Station was -32° C, eighty-proof vodka froze at -27° C.

"I remember," Sergei confirmed as he checked their position on a new Magellan GPS. "Antarctica was a different reality."

"That handheld GPS is *koshachiy zad*." Viktor remarked.

"Stick with English; you need the practice. It's the cat's ass."

"*Da starik*" (yes, old man)." Viktor grinned. Sergei was three years his senior.

Both men were in their forties and retired from the Soviet's Special Operations Force. Three years ago, they'd been stationed at a secret submarine base in the White Sea. Active duty had taken them on two missions to Antarctica. In the intense and constant stress of those missions, what they learned and discovered changed their lives.

"Tell me one more time what General Abramov told you when he asked for your pistol." Viktor leaned forward on his skis and replanted his poles for better balance while standing.

"He told me: 'Hopefully, I won't need this. It depends on what happens at the meeting . . . or after. We're meeting with friendlies, but you never know,'" Sergei readily recalled the general's words.

"And here we are, three years later!" Viktor affirmed; the snow-covered trees reflected in his blue eyes.

"It's been a long three years since Antarctica. Are you still okay with doing this?"

"Still okay, Major."

"There's no rank difference between us anymore; we're done with the army. I'm just Sergei now."

It was a cold night to be illegally departing the Soviet Union. Finland, their destination, was one hundred kilometers west. They knew the Cold War was close to ending after forty-three years but couldn't predict when; it might take another year or more. They couldn't wait any longer; it was time to seek asylum in Finland and recover the Yellow Tablet for General Abramov.

Telemarking downhill again, they welcomed the effortless speed they could develop on such slopes. Then, poetically, they glided on long flat stretches until uphill tundra was encountered demanding double flexing and leg extensions to create glide. Their Norwegian military skis were longer and wider than recreational cross-country skis and did not require waxing. They gripped well in varied terrain and could support a 125-kilo man wearing a twenty-kilo backpack.

The border between the Finland and the Soviet Union (CCCP) is 1,340 kilometers long, and runs mostly south to north. During the Cold War (1947-1991) the socialist dominion of the Soviet Union took extraordinary precautions; nothing was spared in people-containment techniques, which were not meant to keep people out, but keep them in.

Soviet border surveillance began one hundred twenty kilometers from the Finnish border with numerous electronic alarms placed on paths and trails. At sixty kilometers from the border, trip wires could be anywhere, but mostly on or near established routes. At twenty kilometers, a departing Soviet citizen would encounter a substantial barbed wire fence—a modified reindeer fence—with the barbs turned in toward Soviet territory. Both aerial and land reconnaissance operations were conducted day and night.

Supporting these efforts, K-9 (canine) border patrols could be anywhere—they intensified the closer you got to Finland. Soviet citizens caught exiting the CCCP without authorization were guaranteed a one-way train ride to a Siberian re-education camp—unless, of course, they were shot first.

Sergei and Viktor moved fast on the frozen tundra, guided by the luminescent green curtains in the sky and energized by thoughts of living in freedom in Finland. They had prearranged for asylum with help from friends in the Finnish military who they'd met in Antarctica. They would meet them in Inari in northern Lapland after completing the mission for General Abramov further north.

"How far to Finnish border?" Viktor voiced loud enough for Sergei to hear as they kept moving. They had been skiing hard for the last two hours, averaging twenty kilometers per hour, and were tapping into their endurance reserves.

Sergei glanced at his watch then looked at Viktor, reminding him to lower his voice by speaking softly, "Sixty kilometers." Their skis effectively shed the fresh snow with only whispers of sound.

They had avoided the electronic alarms so far. Sergei had the latest location map for them but didn't have one for the trip wires. And the K-9 patrols were unpredictable. Redundancies were the rule when containing a population. Viktor made wide, stem-Christie turns like a downhill skier would before he wedged to a stop. Sergei did the same.

"Water," Viktor said in a dry voice, reaching into his backpack for his insulated canteen. Sergei nodded and grabbed his. Each ex-commando then ate several Reese's Peanut Butter Cups—a necessary energy infusion.

"Great American invention!" Viktor quietly declared, as he devoured one of the 110- calorie chocolate-covered treats.

"Condensed energy—sugar and fat," Sergei responded. They took a short break while looking over the illuminated Russian back-country. The good news: no people and no dogs were in sight.

"Where is the Yellow Tablet buried?" Viktor asked as he adjusted his goggles.

"In Finland, south of the Tripoint."

"The eastern Tripoint?"

"Yes, marked by the Treriksroysa Cairn." The eastern Tripoint is where the borders of Finland, Norway, and Russia meet in the Arctic Circle. It's marked by a mound of stones.

"How far south of it?"

"It's in a sealed canister one hundred meters due south of the cairn—under a rock outcropping where I hid it three years ago," Sergei confirmed.

"I know you told me all this before, it's just my nerves acting up."

"No problem; we're partners." Sergei grabbed Viktor's arm with an affirming squeeze. "Let's go!"

After a series of mostly downhill slopes leveled off, they encountered a stretch of flats infused with pines and spruce where they could glide. Motivated by a desire for freedom and the importance of the Yellow Tablet, they kept going at an energy-demanding pace.

After two hours, they stopped to check the GPS and map again.

"We'll be needing the cable cutter soon," Sergei noted.

"For the reindeer fence?"

"Yes, it will take both of us to cut through it!" The handles of the heavy duty cutter stuck out of Sergei's backpack. "This rusty tool is staying in Russia after it completes its duty."

"I'm ready," Viktor poled away. Wavering green curtains continued to provide perfect lighting in what otherwise would have been a black, moonless winter night. It was 4:00 a.m.

"My boot spray is keeping the dogs away," Viktor grinned through his frozen mustache as they cut a hole in the fence. They'd heard two different patrols with German shepherds barking earlier, but the dogs hadn't scented them.

"I hated to see you destroy that liter of vodka with those odorous skunk thiol chemicals," Sergei quipped. "But one capful on each boot every five kilometers has sure worked. So far we haven't had to shoot any dogs."

"Or comrades! Those dogs have been trained not to chase skunks." Viktor was proud of his pungent concoction.

"We need to keep smelling like skunks until we cross the border. Then I'll neutralize the foul odor with an alkalized-peroxide solution."

"Good. I'm impressed you paid attention in high school chemistry," Sergei mumbled, studying the map once more. "Soon we'll be able to breathe fresh air—in freedom." He folded the map and put it away. They had to ski uphill again, demanding they dig deep into their bodies' energy reserves.

Sergei would always try to think of something pleasant when strenuous exertion was required. At that moment, he thought about his deceased father. His father had been drafted into ski warfare during World War II, in one of eleven Soviet ski battalions that fought the Germans in The Battle of Moscow. He constantly preached the importance of staying fit to his young son. Remembering that wisdom, Sergei vigorously flexed and extended himself. Viktor followed in his tracks. They had to make it to the Tripoint before daybreak.

Helsinki, Finland—February 2003

Finland is a big, happy place with a small population—just five and a half million Finns live in an area only five percent smaller than Germany, which has a population of eighty-three million. Finland is also a cold place, particularly in Lapland where winter temperatures can go well below -18° C (0° F).

This was my third trip to Helsinki; a follow-up to the previous year's visit, when I'd signed up Ivan Popov as my representative in Finland. Ivan, who had taught mathematics at the university in Saint Petersburg earlier in his career, grew up on a farm in Russia, four hundred kilometers north of Saint Petersburg. He understood the needs of both Russian and Finnish farmers who survived by plowing and milking for a living in the sub-arctic north country. He was impressed with the low-cost technology my company had developed for culturing probiotics (beneficial bacteria) on the farm and how they improved silage quality and milk production.

Snow greeted me when I deplaned in Helsinki, early in the afternoon on a cold February Monday. A three-hour flight from London—where I'd had business the previous week—gave me time to get organized for the dinner meeting with Ivan and his friend, Sergei Sokolov.

"Hotel Scandic Hakaniemi," I told the Taxi driver. A nice four-star hotel, walking distance from Market Square in the Hakaniemi district. I unpacked, then went for a short walk to the square's market.

The fresh snow was already packed down from pedestrian traffic and crunched loudly as I entered the old market. I wasn't starved, so a reindeer wrap with a beer was enough to hold me until dinner. The fresh fish aisles in the market were always worth a look. I admired the dozens of northern pike displayed on crushed ice, which were caught in Finland's northern lakes and identical to those I've caught in Canada over the years, where they're simply called "northerns."

Leaving Market Square, I crunched snow over to the World Peace Memorial—a series of bronze statues along an avenue, Hakaniemenranta, that ran parallel to the frigid waters of the Kaisaniemenlahti, which eventually empties into Gulf of Finland. Long, interminable names of streets and places are common in Finland, making a spelling bee more challenging than a calculus exam.

I smiled as I rubbed my hands together and kicked at the fresh snow, happy to be wearing Colorado hiking boots instead of dress shoes.

My thoughts turned to Ivan Popov and our upcoming dinner. My agent in London had introduced us three years before. After our first conversation, I could tell Ivan was a smart man with many talents—one of those men who knows how to make things happen. When I spoke to him on the phone from London on Saturday, I hadn't planned to come to Finland on this trip. He insisted I must. He wanted me to meet Sergei Sokolov, a friend who was in town and, as a commando in the Soviet Special Forces in 1987, had experienced an encounter in Antarctica that I had to hear about.

Ivan told me to meet him for dinner at the Strindberg restaurant in central Helsinki at the intersection of Pohjoisesplanadi Avenue and Korkeavuorenkatu Street. It was across the avenue from a large park, Esplanadin Puisto where Ivan and his wife, Lara, had taken me on a picnic the previous summer. It's considered one of Helsinki's "most-loved green spaces" according to the travel guide—just a frozen blur of white today. I'd arrived early, and the receptionist, recognizing my name on the reservation, had a man named Frans escort me upstairs to a fine, light- filled dining area overlooking the park. "It's a nice view in summer!" Frans said, handing me a menu.

"I know. I was there on a picnic last summer."

"You know Helsinki, then," the tall young man smiled, glancing at my shoes. "Good shoes for hiking in snow."

"They don't quite go with my blue business suit, but comfort trumps style."

"I completely understand. Are you waiting for Ivan?"

"Yes, Frans. Five stars for guessing right."

"Should I quickly bring two Lakka martinis straight up?" Sure proof that the Ivan he knew was Ivan Popov. I laughed. "Yes, by all means."

Ivan arrived fifteen minutes later. When he saw the Lakka martini waiting for him, he exploded with a hearty laugh. I stood up, and we shook hands.

"Thank you for coming. I'm so glad you're here," he said excitedly. "I put in a special request for sub-zero weather—didn't want you to feel homesick."

"It was -18° C in Minneapolis this morning; I feel right at home."

"To winter then, and good Russian vodka flavored with Finnish, wild cloudberry liquor," Ivan held up his martini and proposed a toast. I was definitely back in Finland.

"Sergei Sokolov will be here soon," he checked his watch. We're going to get right into a subject you know something about."

"Let me guess, might that be the alien presence on Earth?" I had figured this out ahead of time. Ivan would often call me in Minneapolis from Helsinki in the middle of the night, wanting to discuss the latest on the alien presence—it was a prime avocation for him. What amazed me most was how it had been kept "secret" since before World War II. Every independent thinker I knew who didn't worry about government or mass-media defamation knew they were here.

"You know it's my number two priority, second only to your probiotic bugs," Ivan assured me.

"I know. Our wives think we're both nuts."

"That's why I didn't invite Lara to join us tonight. I know now there's at least three different hominoid species visiting us!" Ivan reminded me with emphasis.

"Let's see, that would be the EBENS (Extra-Terrestrial Biological Entities), Nordic Blondes, and Reptilians." I acknowledged. There was no need to ease into the conversation. He knew the stories of my encounters in Peru, Brazil, and Australia, where credible evidence pointed to an alien presence. A year ago, I'd mentioned that I knew of three privately funded groups who were quietly investigating the controversial subject. Then I gave him some hints on how to covertly contact them.

"The EBENS are the good guys, the Reptilians our enemy, and the jury's still out on the Nordics. I'm sure there's others too!" Ivan reiterated what he had mentioned to me on the phone several times.

"Where do the little guys with pear-shaped heads and slanted almond eyes fit in—the Greys?" I could see Frans looking at us from across the room. A signal to hold our voices down?

"They're androids: organic robots designed by the EBENS. They're not counted as a species, they don't reproduce."

"So, when any of these aliens are not zipping around in UFOs, they lurk and hide underground, in caves, and other dark places on Earth, right?" I took a drink of the cloudberry-flavored vodka. I wasn't being flippant—Ivan knew that—I was just warming up for a snowy-day discussion on ETs. The subject was certainly of interest to me, but so were many others. With a wife and four children and a business to contend with, it was a challenge to keep up with the aliens.

Frans had seated us away from other occupied tables at Ivan's request, so we wouldn't scare off any regular customers.

We were on Lakka martini number two and snacking on pickled Baltic herring when a tall Russian man walked in. "It's Sergei, right on time," Ivan blurted.

"Greetings, gentlemen." Sergei shook my hand firmly, making direct eye contact, then saluted Ivan and sat down. Ivan pointed to his martini, Sergei shook his head: "*Olut—Sinebychoff*" (Beer) he told a waiter, ordering one of the top brands in Finland.

"I'm pleased to meet you, Mr. Randolf." Sergei radiated a serious military aura.

"Happy to meet you, Mr. Sokolov. When I spoke to Ivan from London; he insisted I come to Helsinki to meet a special friend of his—you."

"Thank you for coming, Mr. Randolf. Ivan and I have known each other since childhood. We both grew up north of Saint Petersburg. I trust anybody he trusts."

"The feeling is mutual, Mr. Sokolov."

"You guys, stop with this Mr. Randolf, Mr. Sokolov formality. Call each other Stan and Sergei," Ivan insisted.

I immediately stood up and raised my martini, Ivan did the same, Sergei raised his beer. We toasted to first names and mutual trust . . ..

"Sergei and I lost contact with each other when he was in the commando business. For about ten years?" Ivan looked at Sergei. "Until he contacted me three years ago at the turn of the century, about the time I met you, Stan.

"Sergei and I are working together with Viktor Petrov, also an ex-commando, following up on a secret meeting they were involved with in Antarctica in 1987. I'll let him fill you in on the details. It's quite a story!" Ivan took a sip of his martini and ate a pickled herring.

Sergei cleared his throat. "Before I start, you men better have another drink, because I'm having another beer." Ivan waved a waiter down and ordered another round, along with more pickled fish.

"First, let me thank you, Stan, for the hints on how to contact the private UFO study groups in Brazil, Peru, and Australia. Ivan requested that I do it."

It was obvious now that Ivan and Sergei had a business relationship in addition to their friendship.

"You're welcome, Sergei. Which contact method did you use?"

"Messages in Christmas cards sent to P.O. boxes in the USA—the ones you gave Ivan."

"Super, that one works well. It's hard for the CIA to read the two billion Christmas cards sent each year in the USA."

Sergei smiled. "That's an order of magnitude better than any covert communication technique the communists taught us."

"Stan's been around the world many times," Ivan added. "He knows a few things."

Sergei then slowly and methodically detailed what he and his comrade Viktor Petrov had observed when Soviet General Abramov and a KGB officer met with two alien beings west of Vostok Station in Antarctica in February of 1987.

"The aliens gave the general a compact computer—a Yellow Tablet—which introduced very advanced mathematics that corrected errors in our quantum math and calculus. Corrections that would allow us to understand how to tap into unlimited energy from empty space—literally. It's called Zero Point Energy (ZPE). Some call it Dark Energy but I was told that's something different. ZPE is clean, infinitely-abundant energy that has been impossible for mankind to unlock and use. The alien's Yellow Tablet explains how to do it, not with just new math but with blueprints for working models. When harnessed and used properly, ZPE does not produce any of the polluting emissions associated with fossil fuels—no hydrocarbon vapors, no sulfur, no nitrates, no CO2." Sergei paused and took a drink of his beer. What he had described sounded too good to be true.

"Where is this Yellow Tablet?" I looked at both of them.

"That was the multi-trillion-dollar question until three years ago," Ivan interjected, then pointed at Sergei to keep going.

Sergei pulled an envelope out of his jacket and handed it to me. "This is an English translation of my notes, originally in Russian, that I made the day after we had contact with the two EBENS. It will add more detail to what I've explained so far."

I eagerly read the notes. His clear descriptions of the craft, the EBENS, and the Yellow Tablet provided additional details.

"Thank you for showing me this. If I had any doubts, they're gone, this nails it!"

"You're welcome!" Sergei declared. "What in my notes convinced you?"

"That EBENS don't wear shoes or boots in Antarctica when it's minus forty! Nobody would just make that up."

Sergei and Ivan laughed out loud. My excitement was peaking—my bum often itched at times like this. I squirmed in the chair.

Sergei took a drink of beer: "Here's what's not in my notes: The general got into a fight with the KGB officer who was unwilling to agree to disclosing the tablet to the world, insisting it must go to Soviet Intelligence at KGB headquarters in Moscow.

The aliens had made it clear to the general that the Yellow Tablet's information must go to the world. General Abramov shot the KGB officer in self-defense during the fight that ensued—with my pistol, on loan to him. I didn't know these 'minor details' at the time. The KGB officer's death had been listed as a suicide and remains that way.

"General Abramov brought the Yellow Tablet back to Moscow in March of 1987, personally delivered it to me, and ordered it hidden in Finland. I did this alone. He had previously briefed Viktor and me on the information it contained, which was enough to seriously motivate both of us. We swore an oath of secrecy and loyalty to the general. That was sixteen years ago!"

"I had expected a discussion on little green men chasing our women around; didn't have any idea it would get this interesting," I interjected.

"Wait until you hear the rest of the story." Ivan repositioned himself and looked around. Nobody was paying any attention to us.

"Okay, Paul Harvey." I grinned.

Sergei took another drink of his beer: "Three years later, one year before the cold war ended, the general asked us to retrieve it. Viktor and I did this in February 1990. We were both retired from the Soviet Special Operations Forces by then and had been planning to leave the Soviet Union. It was easy to decide to help the general while escaping to Finland. We met covertly with the general in Helsinki harbor in March 1990 and returned the Yellow Tablet to him—retrieved by Viktor and me just two weeks before.

"This is where things get really hairy!" Ivan interrupted. "The Tablet got 'lost' for ten years."

"Yes," Sergei acknowledged. "The general delivered the Yellow Tablet to a renowned physics professor in Saint Petersburg in April 1990, a scientist he knew and trusted. The professor's mission was to first understand the equations and explanatory information, then download it onto a PC and laptops. He couldn't do it—the alien math had symbols he couldn't completely understand and the tablet wouldn't download onto a 1990's computer. The professor told the general that the Yellow Tablet's nonlinear equations employed looped infinity functions and differentiations and integrations that involved twists in time and space in five spatial dimensions—overwhelmingly difficult mathematics.

"Some of the exampled solutions involved quantitative calculations that explained how energy coherence could be obtained from random turbulence in high entropy systems, such as fluctuations in the vacuum of empty space—calculations critical to harnessing Zero Point Energy and well beyond the capabilities of our math and primitive operating software at the time."

Frans came by to check if we were ready to order, or if we needed more time.

"What's the house special tonight?" Ivan asked him. None of us had paid any attention to the menu.

"The lamb sirloin pounded with black peppercorns and sprinkled with chopped rosemary and thyme is superb! Very tender. It comes with wild rice from Lapland and a great mix of grilled veggies." The three of us looked at each other, nodding while we quickly scanned other options. Shortly, we all agreed on the lamb.

"Do you have raspberry pavlova with mascarpone mousse for dessert today?" Ivan asked Frans.

"Of course, Mr. Popov. We make it fresh daily." Ivan looked at Sergei and me; I nodded, Sergei thought about it.

"Could a commando eat it in public?" I grinned.

"Only if he's retired!" Sergei replied.

Frans smiled and collected the menus.

With our nutrition details settled, Sergei continued: "So, the professor in Saint Petersburg, with the general's consent, personally delivered the aliens' Yellow Tablet to an exceptionally gifted professor of physics and mathematics in Prague who, enthralled by the tablet's information, disappeared with it in Switzerland for ten years. General Abramov passed away during this time, and I had given up hope the tablet would ever see the light of day again. Then, out of the blue, three years ago, I got a package in the mail from a woman in Moscow—Natasha Abramov!"

"Wow! Let me guess, she was the general's widow and the Yellow Tablet was in the package?" My heart rate skyrocketed. Ivan handed me my martini. I could understand now why Ivan hadn't mentioned this in one of our late-night phone conversations. A private connection in Helsinki was a must.

"Affirmative," Sergei acknowledged. "The Yellow Tablet came with a note from the general addressed to me:

'Major Sokolov, here's the Yellow Tablet with instructions on how to download it, including the type of computer and operating software you must use. The information was never shared beyond the two professors—it will truly advance humanity and protect the environment by making free energy available! Included are some notes from the Czech professor which provide important clarifications, he had to wait for specific advancements in computer design before the tablet's information could be downloaded and transferred—ten years! I will not be around when you read this. Your mission is to get the Yellow Tablet information out to the world. Much Luck, General Abramov.'

"I don't know any more details, Natasha didn't want to speak about it. All she told me was that her husband had instructed her to send the package to me after he died—he'd had cancer. The package was wrapped and addressed to me when he gave it to her. She did not know or care what was in it. *Army stuff*."

"And here we are!" Ivan declared. "The Group in Peru is helping the most. Viktor's in charge of managing this; he's in Lima right now. You can meet him on your next trip; he loves chemistry!

"In just the last eight months we have sent 2003-edition laptops, downloaded with the Yellow Tablet's information and the Czech professor's notes, to eight competent physicists around the world. Their mission: follow the tablet's instructions, prove the validity of its information, then make working models of devices that run on Zero Point Energy (ZPE).

"We've heard back from three of them so far: one each in Finland, Peru, and Australia," Ivan explained, grinning big.

"And they were able to tap into Zero Point Energy?" I asked excitedly, still squirming.

"Yes!" Ivan confirmed, proudly. "With working models."

"This is phenomenal! Earth-changing!" My mind was racing.

"We know very little about ZPE itself, just that the alien math explains how to tap into it," Ivan confessed. "Can you enlighten us beyond what Sergei explained? I know we've got a big cat by the tail, but don't know if it's a lion or a tiger, or what happens if it gets out of its cage?"

"It's likely a hybrid of both cats on supernatural steroids!" I suggested. "Here's what I know about the theory of Zero Point Energy from brain storming with my physicist buddies:

"Zero Point Energy (ZPE) is not nuclear, explosive or radioactive—it would be clean and safe if it could be harnessed. It's the all-pervading energy that defines the fabric of space—the ether. ZPE is intrinsic within the vacuum of space and essentially infinite but limited by its random fluctuations and inability to act predictively with ordinary matter. Einstein called its recognition his biggest mistake and ultimately disregarded it. In actuality, dismissing it was his biggest mistake. He had called it the Cosmological Constant—the ether medium that permitted light to travel through space.

"ZPE is the only energy that remains at absolute zero—or zero degrees Kelvin (-273.15 C, or -459.67 F)—the coldest anything can get. It was previously thought that everything stops moving at zero Kelvin, where all heat is depleted. Current observations disagree with this—indicating there is non-thermal motion at the atomic level due to ZPE. For example, electrons still zip around inside atoms, and quarks continue to vibrate inside protons.

"Here's the big problem: ZPE is said to be impossible to extract and use because it's at maximum entropy (disorder), is incoherent, and vibrates at very high frequencies. However, a recent discovery indicates that regions of order can arise spontaneously in certain high-entropy systems—like with sub-atomic ZPE tornados or vortices akin to real tornados in high-entropy thunderstorms.

"The challenge is to get coherence of trillions of random orderings (ZPE tornados) and collate them by frequency so they can be used as a productive energy source to run generators and produce electricity. Literally, free electricity!

A suitcase-sized *ZPE coherence-collator* will run your car or power your home for decades, centuries, with no polluting emissions . . .." I paused and took a breath.

"So what's the plan to disclose Zero Point Energy to the world without getting killed?" I asked the two of them.

"Let's eat something first, then we'll tell you," Ivan suggested. Frans and another waiter brought out the food.

Postscript 2018

Essentially, any new invention that violates the current scientific paradigm—the world view—is rejected and ignored and its promoters discredited. Ivan, Sergei, and Viktor expected this would happen. The Group in Lima was most helpful in circumventing many of these impediments—they helped develop a conservative, phasing-in plan to discreetly initiate ZPE use on a small scale. Such that, by the time serious funding for expansion would be required, complete proof of efficacy would exist. As of 2014, I was told that a significant number of suitcase-size *ZPE coherence-collators* were running electric generators that produced 100,000 to 250,000 watts (100-250 KW) of direct current per day. Small units, but large enough to operate ten to twenty-five average American homes. A twelve-volt truck battery is used to start the ZPE devices, then disconnected. The generators run perpetually without any inputs other than mechanical maintenance. An undisclosed number of isolated, tribal communities in remote parts of Southeast Asia, Africa, and South America are using these—they're located mostly in dense jungle areas or on secluded islands.

I suspect the EBENS are involved with controlling the scale and timing of interfering. However, the urgency to speed the process up is obvious. Maybe the reason the EBENS are taking so long relates to conflicts they're having with other alien species—those that don't want mankind equipped with free energy. Stay tuned . . ..

Thirteenth century Holy Trinity Church —
Hinton in the Hedges enroute to Stonehenge

Chapter Two

THE STONEHENGE CODE

Mysteries of the Alien Presence

Wiltshire County, England

I was driving on the M-4 highway east of Swindon on a sunny day in early July 1999—London was eighty miles behind me. I thanked the English weather gods for the fine weather; it was always appreciated on the island. The Land Rover I'd rented looked to be overkill; dry weather would be here for the entire week. Thinking I'd be driving off-road in places, like on farmland, I didn't want to risk getting stuck in mud during one of England's legendary rainstorms.

I turned south on the A-4361 in Swindon and headed to Avebury, where I had a reservation at the Avebury Lodge—a quaint bed and breakfast. I managed driving on the left side of the road quite well—I'd done it enough. Downtown London requires full attention as do most roundabouts, but country roads are clear sailing.

The Avebury Lodge was built in 1720, right in the middle of Avebury and its famous stone circle. Multi-ton sarsen stones stand just outside the lodge—a prelude to those at Stonehenge, twenty-seven miles south. These sandstone monsters were brought here from Wales around five thousand years ago but, unlike those at Stonehenge, you could walk up and touch these—no charge. I checked in, then went out into the courtyard to relax among the stones.

As I touched the ancient stones—smooth in some spots, rough in others—I thought about the previous week. My business partner in the U.K., Pat Finnegan and I, had spent the week working with local dairy farmers. I had helped Pat develop a probiotic gel (containing *Lactobacillus acidophilus*) that kept cows and calves happy and healthy by improving their digestion. Preliminary results had been positive: more milk from the cows, fewer calves with scours.

I'd stayed with Pat and his wife Doris for the week at their lovely thatched-roof bungalow in Hinton-in-the-Hedges—population: 179, houses: 39. A delightful place to relax in the Northamptonshire countryside, eleven miles southeast of Banbury.

Next to their home is the thirteenth-century Holy Trinity Church and graveyard. Wikipedia describes it best: "The church is early English; consists of nave, chancel and North aisle, with a low square tower; and contains a remarkable ancient altar tomb, and a very ancient curiously-carved font." My early morning walks usually started around the church.

The walk I had taken the previous morning was one I couldn't stop thinking about. As usual when I stayed with Pat and Doris, I'd be up at daybreak and start walking behind the church "in the hedges." The rapeseed fields were bright yellow and in full bloom in July. Japanese tourists came just to see them. After walking about a half mile, I turned and crossed a hedge that bordered one of the fields. A haze covered the middle of it as the sun was rising. Some distortion in the yellow flowers appeared under the haze, but I couldn't see well with the sun in my eyes.

As I walked on, and the earth repositioned the sun, I could see a large circle shape, maybe fifty yards in diameter. It appeared to have other circles inside of it, all concentric with the outermost circle. I looked around; there was nobody anywhere. It was rare for me to see anyone on these walks. I stepped into the rapeseed bloom and headed toward the circle. The plants were three feet tall and tough—more brittle than barley or wheat. I noticed there was a healthy population of honey bees as I picked one of the flowers. Four, silver-dollar size yellow pedals, similar to scallop shells, formed a square pattern.

I knew what I was looking at now but wasn't ready to admit it yet. It was like when I first saw a kangaroo in the wilds of Australia. I of course knew what it was, but couldn't appreciate *how* it was.

What, conceivably, were these concentric circles doing here? They were in the middle of an expansive rapeseed field, couldn't be seen from the road, and were distant from any buildings. I was so ingrained as an investigative scientist, always seeking more data, that I couldn't appreciate the forest for the trees.

Come on Stan, you're finally seeing one. "Not one, but seven inside of one!" I said out loud. "Seven concentric circles. Crop Circles!"

I'd wished I could have seen them from the air, hovering from a hot air balloon to see details. I looked at the bent stems that formed the edges of the circles. They were not crushed, just bent, and seemed laced together in places. I didn't want to disturb the perfect circles, but I had to know more. What I'd read about crop circles claimed they were hoaxes, made by men with time on their hands and paid for by bars and restaurants to encourage the tourist trade.

But cereologists—crop-circle investigators—claimed they could prove paranormal involvement at least twenty percent of the time: In their testimonies, observers claimed the circles were made at night by colored orbs of light that zipped around just above the fields. Grapefruit-sized spheres of light—white or green, sometimes orange—supposedly utilized plasma vortices to bend crops in mathematically perfect designs.

I didn't have a headache or feel dizzy, as others had reported when walking through crop circles. Didn't see any UFOs or lizard men. No crackling sounds or temperature differences inside the circles versus outside.

Suddenly, one difference became clear: I noticed there were no bees in the circles! In fact, their wings changed pitch to a higher frequency as they approached the circles. Did this mean something . . . magnetic? *Where was my expensive gauss meter that measures magnetic field strength?* At home in Minnesota, of course. My wife reminds me often that I've got every scientific instrument a travelling scientist could want; except rarely do I have the right one with me when I need it!

I recalled the tail on the giant Nazca monkey geoglyph in Peru I'd flown over ten years ago. It twisted into seven concentric circles. One couldn't discern the various Peruvian geoglyphs from the ground, had to do it from the air.

Then I recalled the painted snakes that coiled into seven circles in

35,000 year-old Aboriginal Rock Art in Australia's Northern Territory. And the Anasazi petroglyphs from 600 AD, in the U.S. Southwest, where concentric circles were carved alone on rocks and were not part of an animal. Like here in the rapeseed! What did this all mean?

When I got back to Pat and Doris' bungalow, they were having muffins and tea on their patio. It was a beautiful morning.

"Longer walk than unusual," Pat remarked in his professorial baritone.

"Yes, it certainly was."

"You didn't see our pussy cat, did you?" Doris looked concerned. They had a cat flap on their door, and Snooper would go out late in the day to hunt mice over by the church, but always came back for a cup of milk in the morning.

Before I could say no, Pat said, "Sit down, Stan, and have a muffin." He poured me a cup of tea. "I'll go check out that back door on the church; the cat knows how to open it when it's not locked." Pat stood up, all six foot four of him, took a sip of his tea, then headed for the church. I could picture the cat jumping up at the door handle, relentlessly trying to get both paws around it until it opened.

"I'm sorry I interrupted you, Stan. Please tell me about your walk." Doris smiled. "I worry about that old cat."

I told Doris the story about the crop circles.

"Rare to see them in rapeseed fields; wheat and barley are much easier to bend," she verified. Doris had taught English in secondary school before retiring ten years earlier. She was an avid reader who seemed to know a little something about everything. Smart lady. She took the teapot and filled both of our cups.

"What are you thinking?" I asked her.

"Please tell me again that what you saw was simply seven concentric circles, no other swirls, half-circles, or other complicated geometric designs?"

"Yes, that's right—just seven concentric circles.

"Evenly spaced?"

"Good question, I believe they were." *How could I miss checking that?*

"Then it wasn't Boris and Blake! They're Northamptonshire's infamous crop-circle hoaxers. It was *them.*" She pointed up.

"Codswallop!" Pat had overheard us as he carried Snooper over to her milk.

"Darling, don't get upset. Stan was just explaining what he saw on his walk."

"I'm not upset!" Pat cleared his throat. "I don't want Stanley to start thinking we're a couple of nut-rolls." He broke his muffin and placed a piece of it in the cat's milk.

"Stan, start again, chap. Explain the details of what you saw." I told Pat the story.

"It's too simple a design for Boris and Blake, they're able to do the advanced circle pictograms now—it has taken them ten years to evolve. They wouldn't waste their time just making concentric circles." Doris pointed up again, suggesting other involvement. "I've read that such circles may indicate a meeting place."

"Codswallop!" Pat declared again. "It's likely two young blokes were trying to learn the ropes on how to make crop circles by practicing in rapeseed. In the middle of nowhere so tourists can't critique them."

"Maybe," I interjected. Then I explained what I'd experienced in my travels regarding concentric circles, and that often the simple designs were the most authentic. That did get Patrick to raise his white, bushy eyebrows. Doris filled his teacup.

"The aliens are sending messages to each other." Doris winked at me.

I must have dozed for almost an hour while sitting next to one of Avebury's sacred stones; rethinking that crop-circle experience. Stonehenge was only a short drive away. I'd seen it five years ago from under a wet, wind-blown umbrella on a cold December day. Today was a five-star summer day. With no time to waste, I went back into the lodge and booked a hot air balloon ride for the afternoon. Then I was off to Salisbury in the Rover.

It's hard to explain a balloon ride if you've never taken one. My first was with my son in Africa—over the Serengeti Plain with herds of zebras and wildebeest running below us. That one was followed by a half dozen others, including a very windy outing in New Zealand that almost took us out to sea!

The pilot ignited the burner with a surge of liquid propane sucking in air—the mixture exploded upon ignition with an unmistakable flaming whoosh. Immediately, the hot air lifted all eight of us in the giant balloon basket. We were upwind of Stonehenge about five miles with Salisbury behind us.

The pilot introduced himself as Dexter Quentin and began his well-practiced speech: "Most archaeologist and anthropologist blokes claim Stonehenge survived Noah's flood 4,500 years ago—and was about 500 years old at the time. But the history of this sacred area goes back ten thousand years, according to carbon dating of old wooden totem poles. Back then, a massive stone monument—fifteen times the size of Stonehenge—circled the area. There were lots of circles inside circles."

"Concentric circles with Stonehenge's future location reserved for the very middle?" I said, standing next to Dexter as he ignited more propane.

"There's a chap who knows his onions!" Dexter pointed to me.

The hot air produced by the flaming propane gave us more altitude; the wind cooperated nicely. We headed straight for Stonehenge, which looked like a small stone village in the distance. It was surrounded by green grass for a couple of miles.

"As we get closer, you'll see the mile-wide ring of darker grass that surrounds Stonehenge. It's underlying structure from the Neolithic 10,000 year-old circle," Dexter explained, while grabbing a young girl's hat before it almost blew away.

Wow, this alone is worth the revisit! I thought, smiling at the teenage girl snagging her hat. I never knew Stonehenge was at the center of a much larger, ancient circle complex that came before recorded history.

Dexter continued, "The largest stones—the sarsens—are up to thirty feet tall and weigh twenty-five tons each. They're made of metamorphic sandstone. The smaller stones—the four-tonners—are called the blue stones. The giant, flat sarsen that caps and connects four vertical sarsens, forms three arches called trilithons. Stonehenge has a total of five trilithons. Several had to be re-erected in 1958, requiring the largest construction crane available at the time! How did the ancients lift them up?"

Dexter paused to *whoosh* us up another hundred feet. He had to play the wind at different altitudes to keep the balloon going straight.

"Where did the stones come from?" An Irish gal wearing red Guinness sweats and hair to match shouted out.

"Good question!" Dexter repositioned a few people to balance our weight. "The various 'ologists' debate it. Most believe the stones had to be moved here from Wales a hundred and fifty miles away. Likely by huge teams of oxen or by rafts. At least that's the answer their guesswork produces."

"I believe they got help from the ETs," a teenage boy wearing a Philadelphia Eagles cap suggested loudly.

"Yeah," an older woman agreed. "They probably put the big stones in a secret trilithon somewhere and pressed a key on their computer. And bingo! They popped up in Stonehenge."

"Only a muttonhead would have thought that, much less say it," A stout man who resembled Winston Churchill stammered with an unlit cigar between his teeth. I smiled to myself; Pat Finnegan called ignoramuses "muttonheads."

"Don't light that cigar!" Was Dexter's only response.

"I'll make sure he doesn't!" I asserted, glaring at "Winston."

The man's comment had motivated Dexter to defend the older woman. He revealed it was common these days for guests to connect aliens with Stonehenge.

Whoosh! Up we went again. "Dexter, I think your guests might be getting warm! And not from the propane."

"You might be right, Stan. Just don't tell my boss I went off-license with the speech."

When we got closer to Stonehenge—almost dead center over it to be exact—a large, navy-blue bird landed on Dexter's shoulder—a rook.

"Rooks are in the same genus (*Corvus*) as ravens and crows. They have the intelligence of a Chimpanzee!" Dexter held his shirt pocket open, revealing its contents to the bird.

The rook, using its white bill, pulled out a Payday nut bar from the shirt pocket and presented it to Dexter, who unwrapped it and gave it back to the bird. Holding the nut bar with one claw, the bird then picked the paper wrapper out of Dexter's hand and stuffed it back into his shirt pocket.

"Good job, Mo. Always protect the environment!" Dexter looked the rook in the eye. The bird then glanced around and looked us over, then, still holding the nut bar with one claw, made a series of loud, ratchet-like, clicks.

"He's thanking me and approves of you all," Dexter asserted. I did happen to notice the bird held its gaze on "Winston" the longest. It then turned its head and looked at Dexter with its sky-blue eyes just before it flew away, cawing in celebration of earning the treat.

"Mo's my pet," Dexter exclaimed. "He knows many tricks."

"Amazing bird!" I acknowledged, while Dexter ignited a shot of propane.

"Do you see many crop circles around here?" I waved my arm toward the countryside.

"Saw a good one two weeks ago, west of Salisbury in a wheat field . . . quite a creation. A bunch of circles that looked like planets orbiting the sun in a snowstorm. Big, perfect snowflakes were landing on the planets."

"Who do you think made it?"

"Could have been Boris and Blake. But perfect snowflakes would have been a first for them."

"How do they make crop circles?"

"It's not a secret anymore. The back of their pickup is full of different size lawn rollers and ropes tied to plywood boards."

"Were there any concentric circles in the formation?"

"Yeah, come to think of it. The planet at the tail end—Pluto?—had circles inside it."

"Then the aliens probably made it, or tweaked it!"

"Think so?"

"Cereologists would say they did. Some believe concentric circles denote a meeting place." I leaned forward, rebalancing.

"Don't tell my boss we talked about this! He already thinks I'm a wazzock."

"Is that the big guy who sold me the ticket?" Dexter nodded. "Never fear, my lips are zipped."

We landed shortly after our conversation, right where the pickup bus was parked. I gave Dexter a tip and instructed him to take good care of Mo.

After the balloon ride, I drove back to Stonehenge and purchased a ticket for a walking tour. The tour guide was a Catholic priest who had volunteered to help at the monument during the summer. His name was Father Owen Meyer.

He explained that many of the blue stones had been carried by glaciers closer to Stonehenge than previously thought. Mostly it was the big, twenty-five ton stones, the sarsens, which had to be brought in from Wales.

Father Meyer then explained how at least seventeen shrines formed a large circle which had been recently discovered underground near Stonehenge. And that buried wooden earthworks, cursus monuments, had been erected 5,500 or more years ago, predating the Stonehenge you see today and confirming what Dexter had said.

The priest instructed that, in fact, there was continual evolution within the sacred area of Stonehenge—almost as if new instructions came in frequently to the builders from some unknown source. This took it beyond a cemetery and place of worship into something grander; to a much more powerful *Super-henge,* in the words of some twentieth century scholars. It has been said there is special energy here. I felt lighter, like I had lost twenty pounds. Effects of a paranormal balloon ride?

Father Meyer then explained that Stonehenge previously contained a double circle made from scores of blue stones—possibly competing with existing horseshoe shapes. And that, in particular, experiments with concentric circles continued over many generations, possibly as far back as 7,000 years.

"Many theories exist that attempt to explain the main purpose for Stonehenge, which took over fifteen hundred years to complete in the form you see today. Here's what I believe was the main purpose: An astronomical explanation fits best—it's a megalithic temple aligned with the changing positions of the sun. When a line drawn through the center of Stonehenge, called the avenue, connects with certain buried Cursus monuments, the sun rises on that line during the summer solstice. At sunrise on June twenty-first, the sun's first rays shine through the northeast opening in the henge, illuminating the heart of Stonehenge. Then, on December twenty-first, one hundred eighty degrees opposite, the sun sets on the winter solstice with its last rays streaming through a trilithon." The priest was animated, pointing to the specific locations as he spoke in a commanding voice. The tourist group paid close attention.

He added: "Postholes and other buried shrines in the greater circles surrounding Stonehenge demonstrate dozens of other astronomic alignments in addition to the position of our sun. Some predict lunar eclipses and indicate the location of stars like Sirius at different time periods."

This was, indeed, a more detailed perspective than what Dexter had presented. There were at least thirty people in the tour group, so it was hard to ask questions. I did manage to get to him with one:

"Father Meyer, does the Vatican accept your description of the astronomical purpose of Stonehenge?"

"Yes, actually, it does. The Vatican wants to find truth—not myths. It employs archaeoastronomers who operate an advanced telescope in southern Arizona on Mount Graham. I'm not sure the world is ready to hear what they've discovered."

"Like the alien presence?" I asked him.

"Maybe," he answered. The tour group went silent.

"If there really are aliens, how does that square with Jesus and his messages?" I asked loudly. Everyone was staring at me now.

"Not a problem, if you discount semantics. Aliens, good or bad, translate to angels, good or bad. It's up to us what to call them. Jesus doesn't get distracted by such details. He runs the business of moving souls along a divine path to a wondrous realm."

Hands went up throughout the tour group and a very interesting Q and A ensued. My soul voice offered advice.

"What about souls that are not ready for the path? Do they get another chance down here?" A smartly dressed middle aged woman in a white coat asked as she took off her sunglasses and looked at me. I waved at her, confirming I was a *Homo sapiens.* She smiled.

"Maybe," the priest answered her. More questions were asked—lots more.

"Look what you started." A well attired man in a business suit addressed me.

"I'm Stanley Randolf, a scientist. I travel a lot and like to get to the bottom of things." I reached out to shake his hand.

"Brady Collins. I teach at the University." He shook my hand. We both looked around at the tour group, a picture of animated biodiversity.

As the Q and A continued, Brady and I got to know each other. He had met Pat Finnegan at a university-sponsored lecture one night in Redding.

"We both appreciated McCallum scotch," Brady admitted. "Made it easy to get to know the man—he referred to all other whiskey as codswallop."

"Paddy—my nickname for Pat—loves that word." This led us into a discussion on the crop circle I'd seen. After the tour group broke up, Brady asked if I had any plans for dinner; he said he would enjoy talking more about crop circles. I had no plans for the evening and was happy to accept his offer.

Brady and I agreed to meet at the Market Inn on Butcher Street in Salisbury's old Market Square at 7:30 p.m. It was a fine evening in olde England, and I was happy to see the restaurant had outdoor tables.

"I'm going to tipple an ale," Brady declared upon arrival, taking his suit coat off. I had arrived first and had been seated at good table—one that didn't rock on the cobblestones.

"Sounds civilized; I'll tipple one, too." Unintentionally, I scraped my galvanized chair while pulling it closer to the table. Brady ordered the drinks.

I looked at the dinner menu; he apparently already knew what he wanted. "Any suggestions?" I asked.

"I'm chuffed by their grilled pork sausage, served over fresh garden peas in a bread bowl; goes great with ale. I've had it often."

"Ditto for me then, I'm ready to be chuffed—good word!"

"So, you live in Minnesota but went to the Uni of Wisconsin?"

"Yup, undergrad at Steven's Point and grad school in Milwaukee." We talked about our university days. Shortly, the waiter arrived with two pints of Abbot Ale and we ordered dinner.

"Cheers." Brady held up his pint.

"Cheers back." We both took a drink. I sensed a special energy in this man.

"Want to see some advanced crop circles that were produced in Wilshire County this year?"

"Sure!" I said. He reached into his suitcoat pocket and took out an envelope, opened it, and put a photograph in front of me. It was dated May 2, 1999: A large circle, thirty feet in diameter, was defined by fourteen smaller circles that formed its outer rim. In the middle of it was another circle about fifteen feet in diameter; it had fifteen "flags" surrounding it which aimed at the center point of the circular complex. A forked structure extended out of the large circle with a manicured arrow pointing to yet another circle. Brady gave me time to study it.

"Look carefully," Brady emphasized, pointing at some of the circles.

"Was this in rapeseed?"

"Yes."

I could sense him studying me while I studied the photo. "The circles that form the rim of the large circle have slightly different diameters, and the small flags are either round or square if you look closely," I indicated.

"Yes." Brady gave me a head nod. "Good observations!"

"Are these small differences due to difficulties inherent in rapeseed formations?"

"No," he answered. I paused to think.

"So the slight variations are intentional!"

"Correct, they are subtle codes."

"Hmm. Alien codes?"

"Correct. You're tuned to the right channel now." Brady took a sip of ale.

"Thanks! What I know firsthand about aliens is based on circumstantial evidence I've acquired. Never met an ET . . . not that I'm aware of."

Hearing this, Brady smiled. "Well, permit me to enlighten you with some rather discomforting details: There is an alien war going on right now under our feet. Underground! Good aliens are fighting bad aliens with help from the U.S. and U.K. militaries. It's been going on since the end of World War Two."

"Does it have anything to do with the strange underground explosions that are heard around the world? And unexplainable horn sounds and vibrations?"

"Yes, it does." Brady's face was serious.

"World War Two ended fifty-four years ago. How is it possible they're still fighting? With the technology I presume they have, one side should have vaporized the other by now," I contended.

"You would think so! But it appears they fight on and off in stand-offs."

"Weird." I scraped cobblestones again when I moved my chair. "Maybe they both need to stay around for some cosmic reason, hating each other, like Republicans and Democrats."

"That's an interesting thought! The good ETs, the EBENS, are from the Zeta Reticuli star system. They created the bigheaded, almond-eyed Greys, which are organic droids. Many Greys perish in battles as surrogates for the EBENS.

"The bad aliens are the lizard men, the reptilians that many people joke about—'little green men!' I can assure you they are no joke and they are not little. Reptilians have been here on Earth for a long time. Think 'Antarctica!'"

I just listened. So far, Brady kept his voice down. I could smell food being delivered to other tables where conversations focused on Stonehenge.

Brady continued, "Technology-wise, both sides are close to equal. Our militaries are helping the EBENS. We use tactical nuclear weapons against the lizards when necessary. All underground. Strangely, neither the EBENS nor the reptilians are nuclear armed. There's some rule against it. No ET race is allowed to travel the galaxy with nukes—they punch holes in space-time and screw up the alignments of star gates."

What Brady had explained up to this point agreed with what I'd heard from others. Still, I had a million questions.

He continued, "Here's where crop circle messages come in, it's the one communication method the lizards can't crack. Although not the only option, it's the safest way for EBENS to communicate with each other, and with us, when setting up strategic meetings. Since World War Two, the strategy has been to keep the alien war on an even keel . . . and below ground. That strategy is now on shaky ground, literally." He took another drink of ale and so did I.

"Hoaxing crop circles is part of the game. Boris and Blake and others are paid to create crop circles—the more complex the better. Fill the fields with them, use fractal geometry, new mathematical theorems, fantastic designs. Need to keep tourists happy and the lizards confused. The EBENS, and the U.S. military, can create or modify crop circles with laptop programs that control the orbs: Those bright spheres that zip around above the fields and bend the crops into shape."

Our food arrived. It would have made a French gastronomic connoisseur cry out in joy—simplicity notwithstanding. Both of us bit into one of the toothsome grilled sausages and followed it with a long drink of ale. Dinnertime in olde England.

"Are you chuffed?"

"Yes, and blown away. In my head your information is competing for attention with my hunger command center," I quipped.

We both ate and drank, taking a peaceful pause. I looked around to check and see if we were being surveilled. An old habit I developed during trips to China. Good news: the tourists were preoccupied with food, drink, and stones they had bought.

"What caused hostility between the aliens?" I asked Brady.

"Here's what the EBENS told us: They have supported the human race for thousands of years—ever since Cro-Magnon Man appeared thirty-seven thousand years ago close to the time the Neanderthals became extinct. In fact, they claim they assisted the genetics of the process and have been our guardians ever since. The Reptilians originated in the Orion constellation and have a much longer history on Earth that goes back millions of years. They have long claimed planet Earth but have trouble living on its surface."

"Mr. Brady." I said, munching on a chunk of my bread bowl.

"Yes, Mr. Stanley."

"Here's the sixty-four million quid question:"

"Okay. I'm ready for it," Brady loosened his tie.

"What the Sam Hell are the odds of a Brady Collins meeting a Stanley Randolf in England during a lecture on ancient stones by a

Catholic priest where they agree to have dinner afterwards and discuss nuking lizard men?"

"I'm not a math professor but my guess is they're rather low."

"I agree! Second multi-million quid question: Who can I tell this incredible story to?"

"To anyone you choose! My mission is to get the word out. I need to improve the odds of finding more Stanley Randolfs. Time is short."

"They really let you teach this at the university?"

"I'm sneaky about how I do it—there are powerful financial contributors that support the mission. We've been selective on the students we allow in. No exaggeration; the future of humanity is at stake."

"What subject do you teach?"

"Advanced Studies!"

"That tells me everything I need to know. I'll help any way I can."

"Write stories about this!"

"I can do that."

We both stood up and looked around. The square was busy—many people shuffled by on the twelfth-century streets.

"What's next?" I asked him.

"Well, we still have three more crop-circle photos to examine. . .."

Postscript

If you get the chance to go to England, be sure to visit Stonehenge. And have dinner in Market Square in Salisbury. Enjoy yourself, and don't be afraid to talk to strangers. You might learn something you didn't know—information that may determine your future. And bring along a couple of Payday nut rolls, they could come in handy if a rook lands on your shoulder.

Leopard in high grass — Samburu National Reserve, Kenya
Elephants embracing — Samburu National Reserve

Chapter Three

SAMBURU SAFARI
Inside Wild Africa

Samburu Reserve, Kenya—Late afternoon, October, 1998

"Look at the crocodiles!" my son, Jeff, exclaimed. "Six of them are following each other in the middle of the river . . . moving in our direction."

"Are we high enough?" My sister, Kitty, asked. We were on a wooden deck fifteen feet above a sandy strip of beach that came off the river.

"They can't climb," I assured her.

Two cooks pushed a wheelbarrow full of cattle bones out the backdoor of the Lodge's kitchen and dumped them onto the beach below us. They eyed the location of the crocs, then hurried back inside.

A twelve-footer arrived first, snapping up a large cattle femur, holding it sideways in its mouth. The bone was clean, free of meat scraps, only the marrow would provide any organic nutrition. Nevertheless, the huge reptile ate the entire bone, swallowing large chunks of it—easily a month's worth of calcium for the monster.

"Oh, my god!" my sister shouted, covering her ears. The bone-crunching sounds were loud enough to produce an echo.

Then, a second and third croc arrived, igniting explosive fights that alternated between crunching cattle bones and attacking each other. After the last three crocodiles arrived, reptilian mayhem broke out. They rose up on their legs to gain leverage and clawed over each other to get at the bone pile—right beneath us. . ..

"You'll never see this in a zoo," I remarked.

I could tell my sister still questioned the crocodiles' inability to climb. I reassured her they couldn't, while glancing down—someone had left the gate to the deck stairs open. *Could crocodiles climb a flight of stairs?*

The cooks, now on the deck with us and other guests, threw more bones down to the crocodiles that had angled up on each other, mouths wide open, catching bones in midair. We were witnessing a Jurassic drama starring colossal reptiles that had no need to further evolve. They were here when *Tyrannosaurus rex* roamed the planet sixty-five million years ago . . . imagine an eight-foot *T. Rex* femur in the mouth of a thirty-foot prehistoric croc.

My sister and son had different facial expressions: Kitty's was one of worry, Jeff's, one of astonishment. Both had merit. I would have needed a mirror to see mine— some combination of respect and amazement.

"Don't leave any bones on the deck!" I shouted to the cooks, still eyeing the open gate below. It was 6:30 p.m. and the crocodile-feeding event at Samburu Lodge was well underway. Dusk had arrived.

Earlier that day

It took eight hours to get from Nairobi to Samburu Reserve in the Toyota minivan, a three hundred-fifty kilometer road trip and the beginning of our first week with Peter Kirinyaga, a licensed Kikuyu driver and guide. He was a bright young man whose name meant "A Mountain of Brightness."

Once we got a hundred kilometers north of Nairobi, the drive took us through lush high country then drier, open savannah with mountains in the background. Wet, red-colored roads slowly turned dry and brown as the percentage of iron oxide in the dirt road decreased along with the humidity.

Local people smiled and waved as we passed through villages with open air markets. Young boys sold wood carvings they carried strapped to their bicycles. When we stopped one to check his merchandise, several others quickly appeared.

"Customers in minivan," they yelled to each other in Swahili, then switched to salesmen's English. Africa's "Big Five" (elephant, rhino, Cape buffalo, lion, and leopard) along with giraffes and zebras were well represented in wood. We bought several, making sure they were carved from teak wood so they wouldn't crack. It was fun to bargain with the boys who were pursuing happiness—a sale.

"Safari insurance!" Jeff claimed, holding up a wooden mother giraffe standing over her calf. "Bought it just in case we don't see any in Samburu."

"You'll see many in the reserve—reticulated ones," Peter assured him.

The Samburu National Reserve is more remote compared to Kenya's other reserves. It's a dry African wilderness that depends on the Ewaso Ng'iro River which cuts through its midsection—pronounced *U-aa-so Nyee-ro,* meaning "brown water." It originates in the Kenyan highlands and is the main source of water for Samburu's diverse wildlife. Thick palm groves and dense riverine forests dominate the river's shoreline.

The lioness, Elsa—raised by George and Joy Adamson and made legendary in *Born Free,* the book and movie—along with the lioness Kamunyak, famous for adopting oryx calves, were both from Samburu. The area is also home to the Lokop people, relatives of the Masai.

We had to stop enroute to allow a herd of their goats to cross the road. Halfway across they quit following the queen goat and broke up, surrounding us. Peter, using the horn, slowly inched our way through them. When I looked back, I could see a number of goats had climbed up a dead tree and were standing high on its branches, apparently to put some altitude between them and us.

Ecologically, it became visibly drier as we approached the Samburu Reserve; the tall Hyparrhenia grass progressively lost any hints of green and sparse thorny bushland became more dominant. This was the northern frontier in colonial times.

The effects of uncountable bumps and potholes, which were terrible for the last seventy kilometers, made lasting impressions on our backs and arses, like reproductions of the Toyota logo (imprinted in the seats). The roads would have qualified for Teddy Roosevelt's Rough Riders!

Kitty was five years younger than me and eight inches shorter. Jeff was eighteen, pushing six feet, and in his first year at the University of Minnesota. This was their first African safari, my fourth.

We checked into Larsen's Camp in Samburu with plenty of time before heading to the recommended croc-feeding event at Samburu Lodge. The lodge and camp were in separate locations in the reserve but under the same ownership.

Our large tent was well-secured on a raised wooden platform with two double beds, a cot, mosquito netting, and a fully equipped bathroom. Jeff threw himself onto the cot while Kitty organized the bathroom. After untangling the mossy netting, I dumped my safari bag out on one of the beds and began sorting things out: several guns, long knives, machete, cattle prod, spotlight, magnum pepper spray, etc.—*just kidding!* In reality: Insect repellent, flashlight, whistle, suntan lotion, Swiss army knife, first-aid kit, binoculars, magnifying glass, and notebook.

Charles, the assistant manager, came to our tent to check if everything met with our approval.

"So far, everything looks ship-shape," I told Charles. "First class." I'd been to some questionable safari camps in the past.

"Thank you, thank you, Mr. Randolf. It's our pleasure to serve you. There are just a few simple rules, please. One: don't feed any animals. Two: don't go outside camp unless accompanied by an authorized guard or guide. Three: don't swim or go near the river."

"Is the river dangerous?" Kitty asked him.

"The river is fast moving, muddy, and full of critters that can either eat or poison you," he said, answering my sister, whose face quickly morphed from a smile into a portrait of fear.

"An electric fence surrounds camp for your protection, except along the river shoreline. Call me if you need anything."

"Thanks, Charles, I'll change from my swimming suit to safari pants now," Jeff, the comic, responded with a grin.

Charles grinned back. "Enjoy your stay with us!"

There was a large treehouse for guests not far from our tent. We climbed the ladder up to it, joining an older English couple. The expansive view of the riverine forest and river included six reticulated giraffes feeding on acacia trees. The air was clear and exceptionally fresh—together with the view, it was Africa at its finest.

"They're different from the giraffes we saw last week," Kitty noted.

"Those were Masai giraffes with spots that resembled oak leaves, these reticulated giraffes have connecting circles or polygons, and are about two feet larger," I explained.

"Taller," Jeff corrected me. "The males can reach eighteen feet! Read about them on the KLM flight from Amsterdam." I was impressed; I thought he'd been reading Spiderman comics.

"We heard a lion roar last night. There was no mistaking it," the English woman said, looking away from the giraffes. "The headwaiter told us it was five kilometers from here, well outside the electric fence."

Her husband nodded. "When a lion is only one kilometer away it sounds like it's just outside your tent." Kitty asked how many miles that was.

Early morning—the next day

It was a bright, sunny morning, and a significant number of Samburu's three hundred and fifty bird species loudly communicated with each other as Jeff and I finished breakfast—customized omelets made to our specifications.

"*Jambo*!" We hailed Peter as he walked up to us. Standing tall and gleaming with an ebony shine, he was ready for the week with us. I enjoyed the shortage of indefinite articles in his concise speech, the "a" and "the" were mostly unused. He was a smart man who didn't use unnecessary words.

"Jambo, Mr. Stan, Mr. Jeff," he replied with a big smile. "Where's Ms. Kitty?"

"She's back at the tent trying to decide which of her deodorants will repel crocs." Jeff explained.

Peter laughed. "We'll see them in river with hippos. Maybe see hippo fight croc!"

"It rained last night with strong wind gusts; those tall Petticoat palms were smacking our tent," I remarked. "And one of the fan palms was split down the middle."

"Everything was dry this morning," Jeff added.

"Short rains of October. Wet disappears fast here," Peter acknowledged.

Jeff aimed his Minolta SLR camera at a black-faced monkey ten feet from us . . . just before a waiter took a broom to it. I knew from previous safaris it was a vervet monkey—cautious and smart. They're mostly gray with charcoal-black faces.

"He's scouting for his comrades down by river," Peter explained. "Keep tent closed tight! They steal anything they can carry."

"We know! Jeff learned this in Masai Mara last week, after losing his Minnesota Gopher's cap to one. The monkey was later seen in a tree, wearing it sideways on his head. The cap was irretrievable, despite our offering bananas in trade."

"You know then," Peter replied. "Black -face monkeys are smart! They count how many guests are in a tent. When last guest leaves, if tent not closed tight, in they go."

"Those silly monkeys are everywhere," Peter added, while opening a map of the Samburu reserve showing its myriad roads and game trails.

"Looks like a big, wild place," Jeff commented, shouldering his camera.

"Indeed. We'll explore it—one hundred sixty-five square kilometers. Wilderness extends outside park, up to Ethiopia. Are we ready?"

"I'll go round up Kitty and secure the tent," Jeff volunteered.

"Excellent," Peter responded. "Your father and I will be in minivan." It was 7:00 a.m.

Kitty, sitting in back, wanted to see lions provided she could keep her windows closed! Jeff and I were standing up in the open sunroof, so we didn't mind. Peter had his driver's window open. To me, there seemed to be a greater margin for safety in a minivan than in a fully open Land Rover like we'd had in Masai Mara. A five-hundred-pound lion wouldn't fit through a sunroof—though I had read about one in Namibia that tried.

Peter must have been reading my mind regarding wildlife hazards. "Leopards can be a worry," he commented. Leopards less than half size of lions."

"How so?" Kitty asked nervously.

"They're unpredictable: Two years ago, this French couple was aggravating two leopards on honeymoon—female in heat. Two-hundred-pound male get angry. He jump on minivan; come in sunroof!"

"Oh, my god," Kitty stammered. "Don't tell me this!"

"Kitty, it's okay. We're not going to aggravate any leopards; go ahead, Peter," I interjected. Jeff gave her some water to drink, thinking a double martini might help her more.

Peter explained they had been in a remote area north of here that's now off-limits to safaris; the cats were not people-acclimated. His friend Jacob was their guide.

"How bad did it get?" I asked.

"Very bad! All five people in van wounded. Jacob lose one eye."

My sister was white-faced with fear.

"Kitty, we're not going to aggravate any leopards," Jeff assured her, repeating what I'd said. I knew some people experienced "first safari jitters." Reading about these things in a book was one thing, hearing them firsthand from a guide while in the African bush was another.

"What happened next?" Jeff asked Peter.

"Jacob run out of van and open side door. He had one of those air-can horns . . . he pointed it at male leopard. The loud screeching horn get male leopard to leave and go back to female." I made a mental note: *Loud horns seem to dissuade both goats and leopards.*

"Then what did they do?" I asked. I could tell sharing the full story was starting to wear on Peter.

"They had no radio, everyone bleeding from being clawed. Jacob apply first aid best he could, then drive to Samburu Lodge. Lodge call army helicopter to take them to Nairobi. Jacob lost left eye," Peter lamented. "Very rare for this to happen!"

Peter continued driving. In an effort to lighten things up on our nerves, he decided to switch attention to our bums by discussing the condition of the road. "Minivan preferred on rutted, dry roads in Samburu; Land Rovers with big tires better in Masai Mara where it's more wet," Peter clarified, happy to change subjects. As he drove on the potholes in the road increased in both number and depth. We could hear and feel when the Toyota's leaf springs maxed out and the frame of the vehicle hit the road. While standing up, Jeff and I had to lean against each other for support. Finally, Peter had to slow down and drive partially off-road and push through brown grass to smooth things out—somewhat. Kitty bounced around in the rear seat. The bumps didn't seem to bother her.

Several white-headed vultures flew over a decomposed kill about a hundred yards to our right, squawking with delight. Giraffes fed on the tree line beyond the birds, close to a small herd of gemsbok (oryx)—easily identified by their two, straight, yard-long horns. Good morning, Samburu! The air was now alive with elephant-dung vapor as we drove over a plurality of twenty-inch diameter, circular footprints.

Then, as we approached the river, we saw the elephants, lots of them. A sizable herd was crossing the Ewaso Ng'iro River. Quite a picture—Jeff and I took several. They were going, not coming.

"They're mostly brown, all covered with mud," Kitty observed. A few were red from iron oxide-rich mud. Red elephants . . . first time for me!

"And they're washing each other as they go, spraying the one in front of them without stopping," Jeff remarked.

"Their trunks hold so much water!" Kitty added, smiling now. The elephants at the front of the line were gray again after crossing the river. Once across, several of the teenage pachyderms got out of line and began giving each other exuberant trunk hugs. I had my Nikon N-70 on a bean bag zoomed to 600-millimeters with a 2X converter. Got good pics.

"I've heard they can read your mind," I said, checking for Peter's response.

"They can," was Peter's quick answer. "If big bull senses he's threatened, he can charge. Puts head down and shakes it as warning, then charges, ordering herd to follow. Elephant hunters get trampled before they're ready to shoot—elephants read their minds." I had read about an elephant hunting camp in Zimbabwe that got trampled the night before their scheduled hunt. The elephants must have read the minds of the hunters dreaming about killing them.

"We need to smile at them and think happy thoughts," Jeff deduced.

"Maybe offer some peanuts," my sister suggested.

"Out here, you would need five pounds of peanuts for each elephant—they can't taste anything less," Peter claimed. "They normally eat four-hundred pounds of grass and leaves a day—big bulls eat six-hundred pounds."

I had remembered reading about a bull elephant in Kenya's Tsavo Reserve that charged a minivan full of people and tipped it over, but decided not to bring this up. For me, most adventures presented some danger and could be exciting in direct proportion to the caliber of the danger. On this trip I had a big responsibility: the safety of my son and sister. I needed to keep things calm, cool, and collected as much as possible. No elephant riding, no swimming with crocs, no petting lions.

After the elephants we visited a small lagoon, the size of two basketball courts, which had separated from the river. Aquatic weeds were plentiful and I counted twelve floating heads—pink-eared, puff-eyed, and wide snouted. Below the heads were tons of submerged flesh—probably averaging two tons each. Three additional hippos were out of the water on the opposite shore, swinging their tails.

Peter had to keep us back twenty yards on dry ground because of the minivan's ineptitude in wet sand. Kitty was fine with that.

"A Land Rover could have driven into the water with them," Jeff declared. "For better photos."

Peter smiled, "But only once! Hippos have beastly temperaments—very dangerous! They kill more locals in East Africa than lions." Hearing that, Kitty lost any vestige of a smile.

The air was filled with hippo-dung vapors. Each animal produces around *four hundred pounds* of feces a day, which generates a lot of vapor! The males fling dung by spinning their tail during bowel movements—which are frequent. Apparently, some skill is involved in achieving substantial distance. Good timing is most likely required.

Flung dung impresses female hippos beyond what my words can express. Seriously.

"Is it true that an adult male hippo can bite a ten-foot crocodile in half?" I asked Peter.

"I've seen it," he answered. "The crocodile went crazy in hippo's mouth, twisting and tearing itself apart on the huge tusks. Hippo keep mouth mostly closed until croc saw itself in half—then spit it out." Even though Hippopotamuses are vegetarians that only eat weeds and grass, they're born with a bad disposition and can easily become violent."

"I heard they can bite your head off!" Jeff exclaimed, being serious now.

"Many locals have lost their heads; not happy way to die," Peter acknowledged. "Hippos constantly keep teeth sharp by grinding. Use tusks as weapons for attack and defense, canine teeth for eating. If hippo gets close and yawns, look out . . . he's ready to attack. He can run forty kilometers per hour!"

Two of the males buoyed themselves up in the water and opened their mouths facing each other. Very dramatic. You could have stood a baseball bat straight up in each of their mouths.

"They're threatening each other for territory: control of lagoon and females," Peter explained. "What you see is bluster; young males not fully established yet."

We moved on and the air cleared. Kitty couldn't get dung flinging out of her head. Safaris can reveal surprising information.

We drove on, passing a variety of dry-country grazing animals that further confirmed Samburu as a dry wilderness deep in East Africa: Oryx, Grevy's zebra, gerenuk gazelle, thousand-pound-plus eland and Cape buffalo were abundant. It had warmed up to 30 Celsius (86 F) and was dry and dusty. Rocks from Time's Basement were scattered throughout this extension of The Great Rift Valley, sharing space with thorn bushes and a gray, sage-like plant. Peter turned in the direction of the river once again where we could see green trees and the fertile ecotone that formed between savanna and trees. There was tall grass, mostly brown, that could easily hide predator cats. All three of Africa's big cats—lions, leopards, and cheetahs—hunt in

Samburu.

"There's a pecking order between big cat species," Peter explained. "Lion can steal leopard's kill, leopard can steal cheetah's kill—with no retaliation."

"A cat needs to know its place in feline society. Size matters!" I professed.

But we hadn't seen any big cats yet. We headed back to camp for lunch.

Kitty, Jeff, and I had Tusker beers in the bar tent while the staff set the outdoor tables. Jacos, the bartender, pointed out gecko lizards that had learned to climb the bar's smooth, steel light poles to feast on a consortium of African bugs—sticky tongues were their secret. They preferred flies over beetles.

Vervet monkeys, a genet cat in a tree, and an Audubon field guide of birds had positioned themselves for the lunch event. "Do Not Feed the Animals" was well-posted in several languages.

Eighteen guests were seated outside at wooden tables covered with white tablecloths. We were the only Americans as far as I could tell.

Well-dressed servers brought out the food: grilled prawns from Mombasa, lentils in mushroom sauce over long-grain African rice, and fried tilapia from Lake Victoria, all supported by an amazing buffet of local veggies and fruits. The staff did a good job of keeping the monkeys at bay; the birds were a bigger challenge.

Larsen's Camp bordered one side of the river about fifty yards from us. On the opposite side was wild Africa, where we'd seen a twelve-foot crocodile sunning itself this morning not far from two marabou storks eating a dead monitor lizard—a big one. To some undefined extent, the camp staff wanted us to believe there was a difference and that we were on the safe side of the river. I knew better but didn't want to upset Kitty by discussing it. Maintaining awareness was key . . . and snakes were certainly something to watch for—like lethal green mambas that hid in green trees.

After lunch we drove to the Samburu Lodge, where we had watched the crocodiles eat bones the previous night. The lodge's swimming pool was the attraction today. It was a sizable pool, and the water was clean and comfortably cool. Beyond the pool was a desert garden featuring various African cacti, including a small grove of candelabra (cactus) trees. Kitty did some sunbathing while Jeff and I took a swim.

Large red and blue iguanas sunned themselves on nearby rocks. One was swimming in the pool, which kept several European ladies out of the water. The foot-long lizards had remarkable red-blue contrast, but were outdone by a dozen superb starlings (their common name) flying in and out of a lawn sprinkler. These robin-size, iridescent-blue birds, were gregarious and people friendly. However, whenever one of the iguanas got near the sprinkler, the male starlings would squawk loud warnings, alerting the females of potential danger. I'd seen these outstanding birds on all four of my African safaris.

I ordered rum daiquiris from the pool bar for Kitty and me and a Tusker beer for Jeff. It was time to relax and rehabilitate our bums.

The afternoon game drive started at 4:30 p.m. Jeff and I assumed our standing-up positions in the sunroof, while Kitty sat safely in the backseat with her windows closed.

"Let's go find lions!" Peter said loudly.

"We're all for that," Jeff acknowledged as we drove out the gate onto the brown dirt road we had taken before lunch.

We'd hardly gone a hundred and fifty yards, where we had seen nothing this morning, when we surprised two mature lionesses walking the road in front of us. Peter slowed down.

"They're stalking those Impala," Peter said, pointing to the herd on the savannah. Tall grass bordering the road hid the lions from the impala.

Kitty clapped and screamed, excitedly. At that, both cats glanced back at the minivan and snarled.

"These lions aren't as people-conditioned as those in parks closer to Nairobi. They have many places up here with no roads or people—undisturbed bush country," Peter informed us.

"Camp is so close!" Kitty said, nervously.

"Electric fence keep them out of camp," Peter affirmed.

"And they're going in the opposite direction of camp," Jeff added.

I could tell Peter was concerned by the lion's snarling—normally, lions completely ignore safari vehicles. Most grow up with them as a routine part of their lives—people in the vehicle are a natural part of it and recognized as such by the big cats. Get out of a vehicle, and you become unnatural and strange—an easy meal!

When our minivan got closer to the lions—just six or seven feet away—they turned their heads toward us and snarled again, louder. Then, the one closest growled with deep resonance—Jeff and I quickly sat down. Peter closed his window until we passed them. This was the first pair of lionesses to both snarl and growl at me in my safari history. Usually that was left up to the males.

"Maybe they're agitated because we upset their stalk?" Jeff commented. We could see the impalas running now, some leaping with all four legs off the ground.

"Maybe," I responded. These two cats were big, with paws the size of dinner plates. Their bodies wouldn't fit through the sunroof, but their paws could—with lethal claws that could really stir things up in here. I stood up and pulled the sunroof shut.

"In case in rains," I said. My sister wasn't amused. We drove on looking for less agitated lions.

Jeff and I stood up again, with cameras ready. An hour passed as we bumped along; Peter was driving in ruts created by other minivans, leaving a trail of fine brown dust. In the distance, a herd of Grevy's zebras grazed in a depression that was damp from last night's rain. Through binoculars I noticed they were mostly eating wet mud—for moisture and minerals? Hmm, certainly not for calories.

"Those white-striped zebras, *Equus grevyi,* are largest living equids—stallions can reach nine hundred pounds," Peter explained. "They're endangered, less common than Burchell's or plains zebras, *Equus quagga,* which have thick black stripes." It took zebras with white pinstripes to ignite Peter's zoological grandiloquence.

There were fifteen zebras in the group. One was apart from the others, the sentry on guard duty. "Young zebra favorite lion food," Peter added. "They always watch for lions."

Several miles past the zebras we turned onto a road that paralleled the river, and just beyond the dirt intersection, saw two male lions partially hidden in tall grass.

"My god, they're huge." Kitty squeaked out her words.

"Big Leo and Young Leo," Peter whispered, introducing us to the two male lions. He knew them. Big Leo was at least nine feet long and weighed over five-hundred pounds. Both cats had obsidian-black manes encircling their yellow-bearded heads. The lions were a darker tawny color than the blonde ones in wetter climates like Masai Mara.

Jeff and I clicked away with our Nikons and Minoltas, just as Big Leo let out a deafening roar—followed by a staccato of air gulps. At an estimated 115 decibels, twenty-five times that of a power mower, it almost knocked us over. Jeff and I sat down quickly.

"We surprised them." Peter kept the van idling in neutral. The commotion had awakened Young Leo, all eight feet of him. He sat up and stared us down, then yawned. We watched quietly . . ..

"Young Leo is Big Leo's cousin," Peter explained. "Their pride is somewhere around here. Five weeks ago, there were four lionesses and ten cubs with the two Leos. Unlike prides where male lions kill cubs—infanticide to get females back in heat—two Leos protected their pride's cubs." Peter paused, giving us time to think about that.

Peter continued, "One month ago, big Cape buffalo accidently step on two cubs, crushing them. Lions rarely attack adult buffalo. Two Leos not around when cubs stepped on; lionesses told them to kill that Cape buffalo. They knew exactly which buffalo—twelve hundred-pound bull. They worked together in kill-or-be-killed fight. Cape buffalo bulls very dangerous; they have heavy, very sharp horns and attack by ramming and shaking their head. I saw one gore large bull rhinoceros to death!" It was clear Peter had experienced part of this story.

He continued, "The two Leos kill this Cape buffalo. I think they run it in circle until tired. Then Big Leo attacked rear quarter first, biting into it and holding on with its claws deep in buffalo so Young Leo could attack neck. After it was killed, they didn't eat buffalo's tasty meat, but left carcass for hyenas and vultures. This caused entire buffalo herd to move twenty kilometers."

"More than instinct was at play." I looked at Peter. The movie *The Ghost and the Darkness* with Michael Douglas and Val Kilmer, tells of two male lions that had learned to work together synergistically—but not with a benevolent motive. Together they stalked and killed Indian railroad workers in Kenya's Tsavo Reserve during colonial times. A true story. "There are lions and there are lions."

"In animal kingdom, smart ones work together," Peter added.

We drove on and located the lion pride about a mile from the two Leos and spent an hour watching the eight cubs play with each other and the four lionesses—it was hard to decide which cubs belonged to which lioness. They alternated their affections. The cubs that weren't rolling and pawing on the adults were scrapping with each other over a piece of zebra hide. I could sense they were sharing the joy of being alive.

"They've got each other and the world by the tail," Jeff declared.

"Until another Cape buffalo stumbles by," Kitty supposed, uttering her first words in a while. She was beginning to adjust to the day's adventure, its reality, and its stories. I gave her a big hug.

I knew a philosophy professor who would have been tickled pink by Jeff's and Kitty's contrasting comments. He called those moments of joy and happiness—those fleeting good times that ultimately come to an end—the "golden twilight zone." The two lion cubs that were stepped on spent only a short time in that zone. Sadly, so do many people.

As we drove further, closer now to the river, a regiment of helmeted Guinea fowl appeared ahead of us on the road. They were eating worms out of elephant footprints that were enriched with manure—a variety of vermin, including fly larvae, ticks, and locusts augmented the worms. These plump, red-billed, blue-necked birds live in dark indigo bodies spackled with white spots and have what looks like a bent index finger for a helmet. Endemic to Africa, they play an important role in pest control. The males attract females by squawking like a cheap New Year's Eve noise maker.

"They look like purple turkeys," Kitty observed.

"Are they edible?" Jeff asked.

"You would say the wild ones taste gamey. Those raised on farms and fed corn and beans taste good," Peter replied.

It was dark at 7:00 p.m.—the transition from dusk to night happens fast in Kenya; both predator and prey must adjust fast. The blackness of Africa is a different world. The boundary between dusk and black-dark is what the guides refer to as "last-light." It's a special time.

The male leopard was back in a favorite broad acacia tree. Its kill, a long neck gerenuk gazelle, hung on the branch above it. Peter had saved this for last.

At five years of age, this leopard was a marvelous specimen: large black rosettes formed patterns on its golden back; solid black spots dominated its legs and belly.

There was enough moonlight to see it as Peter circled the tree, twenty-five yards away. At the sound of the minivan, the leopard stood up. With the moon at its back, it cast a giant cat shadow in front of us. We were alone with him.

"Amazing!" was Jeff's one-word description.

Kitty was squeezing several Kleenex tissues tightly, surely thinking about Peter's story of the French couple.

"He's a big one, over two-hundred pounds," Peter estimated. "That gerenuk with broken neck will feed him for two weeks if nothing steals it. Martial eagles and baboons eat leopard's kill when cat's away. Hyenas do too, but they can't fly or climb trees."

"He's really big," Kitty mumbled, peering up at it through the back window. She squeezed her tissue clutch again.

Jeff and I were standing up in the sunroof taking turns with the binoculars. Peter shut the motor off, which got the leopard to lie back down on the acacia tree's substantial, horizontal branch. Now, *this* was Africa! My soul voice sang.

We got back to camp at 7:30 p.m. in time for dinner at eight. Jeff and I went straight to the bar tent, while Kitty went to our tent to freshen up. We ordered Tusker beers while checking on the pole-climbing geckos. Their tongues were actively attacking bugs—one was trying to eat an enormous fly.

The bar tent had an open area out back where other guests were sitting. We went out with our beers and found a table. There was something going on over by a Tiki Bar sign; thought I was in Key West for a moment. The sign had a spotlight on it and had attracted an impressive collection of large moths—some the *size of my hand*.

"Here's to a bunch of magnum moths." Jeff lifted his beer. I did the same and we toasted the Lepidopterans. Some looked like owl heads with big, silvery-white eye spots.

"Protective mimicry," Jeff noted. "Anything afraid of owls won't mess with them." He had learned a bit of entomology from his dad.

Suddenly a large male hornbill, apparently not afraid of owls, swooped down and picked off the largest of the owl-imitating moths and flew with it to a nearby gum tree where a female hornbill was waiting. He held the moth by its wings allowing her to snap into its fat abdomen, then she gulped it down without the wings. The male flew back to the sign to get her another moth—this time one that looked more like an F-15 fighter jet.

"Wow!" Jeff said. "It's good to be a female hornbill."

The tables were set for a dinner under the stars. Kitty was at ours, well into a Bombay-Sapphire martini, talking to a couple from Spain when I sat down.

"I'm relaxing, Stanley," she held up her drink.

"I'll drink to that." I took a slug of my Tusker. Jeff was in the bar tent's porta potty.

Before I could mention the hornbills, Kitty pointed to the genet, in the same tree as this afternoon, only awake now.

"They're a distant relative to house cats; they eat birds and rodents and are nocturnal," Jeff said when he returned. He had been reading the African wildlife guide I gave him.

"Reading in a porta potty is bad for your lungs," I advised him.

"It's a compromise; I do it for the sake of higher education."

We all laughed, including the couple from Spain and the headwaiter.

The genet was unique. About the size of a fat domestic cat, it had a long, striped raccoon tail attached to a tawny, black-spotted body with big ears and a foxy snout—an African carnivore that evolved during the Pliocene Epoch *five million years ago*.

"That's Ginny, our pet," The headwaiter said, as he brought out smoked beef and consommé soup for starters. "Don't let her steal any food; she'll get table scraps later." It made me recall a pet genet in South Africa. I was staying with business friends at their home in Johannesburg when it crawled in bed with me in the middle of the night. I had been sound asleep until it dragged its three foot tail between my legs and stuck its cold, wet snout into my armpit. I literally skyrocketed out of bed shouting something in Afrikaans. Not funny back then!

The enticing aromas got Ginny to come to attention. Eland steaks in ginger sauce with carrots and French fries were brought out for the main course. We had already sampled the fabulous buffet of local veggies and fruits.

"*Hapana! Hapana!*" (No! No!), one of the lady servers shouted at Ginny in Swahili, pointing at her, standing up now, in the tree. "*Kukaa, Kukaa,*" (Stay, stay), she voiced with authority. This got my sister to finish her martini and Ginny to lie back down.

European-style flan was for dessert. After dinner we were serenaded by Larsen's service ladies—the cooks and maids. They sang deep, haunting, African melodies with no instrumental accompaniment. The harmony of their voices was remarkable—soul penetrating. My soul voice celebrated.

After that, it was time to call it a day. The five young stars comprising the Southern Cross shone down on us as we walked back to our tent. These ten-to-twenty million year old stars, infants by comparison to our sun's 4.6 billion years, have followed me on many journeys into the southern hemisphere.

The camp's electricity shut off at midnight, and the pole lights went dark. A propane lantern, the only light, flickered over by the bar tent. I stood out on our tent's deck appreciating the darkness at 1:00 a.m. October's short rains which produced less than 0.1 inch of moisture per storm had obscured the stars; at least we'd seen the cross earlier. The storms of October in Samburu were largely wind.

Jeff had passed out on his cot inside. Kitty was tinkering around in the bathroom, she couldn't sleep. I was enjoying a Crown Royal whiskey in my coffee mug when a flashlight beam flashed by me. I pressed the button on my Mag-Light and flashed back. A man, armed with an automatic rifle, walked over to me. When he got close, I recognized his Kenyan Wildlife Service (KWS) uniform.

"What's happening, officer?" I asked in a low volume.

"Didn't anyone warn you?" He was obviously surprised to see a guest outside at this hour.

"No, about what?"

"Two female lions were seen just outside camp tonight," the officer explained.

"We're protected by the electric fence, right?"

"The electric fence is down, shorted out by a large monitor lizard."

"Oh, my god!" Kitty exclaimed, sticking her head out the tent's door flap. She had overheard us.

"Go back to bed, young lady," the ranger told her. "*Hakuna matata;* there's nothing to worry about. We've got the matter under control." He pulled a notebook out of his pocket, paged through it, found our tent number and my name, then said to me: "Mr. Randolf, you need to go to bed, too."

Visions of the two snarling lionesses from this afternoon flooded my gray matter. Kitty was sitting up in her bed unable to sleep. Jeff was snoring, somewhere in the Dreamtime. I laid down on the bed, pulled the mosquito netting over me, and tried to fall asleep. I hated the netting, it tickled my feet.

The ranger must have turned the camp lights on. Half asleep now, I could see shadows on the side of the canvas tent—full bodied shadows of something moving. I nervously got up and peeked out the door flap. There was some wind now . . .. It was hard to get a good view from the flap. Two hunched shadows side by side were off to the right! They were moving back and forth. *Lion shadows? Jesus! Where's my elk hunting rifle? At home in Minnesota, of course.*

Kitty had finally fallen asleep; Jeff was still dreaming. I had to get a better look.

The bathroom has a window! I remembered. I crept in quietly and peered out the window . . .. I could see now with a wider view. What was I looking at? *It's that split fan palm bending in the wind, you idiot!* I scolded myself. The shadows of that broken palm had looked like two lions walking together.

My imagination easily fools me at times like this. At 2:00 a.m. I went back to bed, cursing the mossy netting.

I woke up at 3:30 a.m. to Kitty's shouts from the bathroom. "Spider, giant spider!"

I jumped up, grabbed my Mag-Light and ran into the bathroom. Why did it have to be a spider? I hate spiders. I felt like Indiana Jones when he was confronted by a pit full of snakes in *Raiders of the Lost Ark*. He hated snakes. *But this would only be one spider*, I told myself.

"Where?" I shouted, which woke Jeff up.

"Under the toilet." She said, trembling.

"Get my coffee mug, it's on the deck," I told Jeff. I could see the spider—an ugly, orange-black critter with a body the size of a silver dollar. It had its two front legs in the air, probably trying to smell me or see me? Eight of me, one for each of its eyes.

"There's booze in the mug," Jeff said, sniffing it.

"Hand it to me," I grabbed it and swallowed the two ounces of Crown Royal whisky in one gulp; then attempted to capture the large arachnid by putting my mug over it. Crushing it was not a good option; I had boot-kicked a large dock spider in Canada once and green goo shot out of it all over my leg! So the mug had to perform for me and catch this one alive. I missed on the first try, which invigorated it, big-time. Moving fast now, the spider was coming toward me—bad news. Usually bugs and spiders run away from danger!

I tried again and somehow was very lucky—got it—whew! Most of its eight legs and body were under the mug on the floor. Now what?

Jeff had found a large 5x7 inch post card of Mount Kilimanjaro in his suitcase and gave it to me. I waved it around to check its stiffness before trying to slide it under the mug.

"Let me do it, Dad," he advised. I happily handed him the card and before I could stand up, he had the bug in the mug and was out of the tent with it. Kitty screamed.

"Did you kill it?" she asked when he came back in.

"I let it go!" he said proudly. It happily jumped away heading toward the river.

Now, it was certainly time to call it a day, er, a night, once again. In the morning Peter would take us farther north into black rhino country. There were not many of them left. Then on a search for the elusive Kudu antelope and, of course, more cats.

Kitty wanted to see reticulated giraffes up close plus other grazing animals—she had been paging through Jeff's wildlife guide.

"And maybe an ostrich or two," she said, before falling asleep . . ..

Postscript:

I love Africa and have learned to respect it. There is beauty and excitement but also danger. Being cautious is not unmanly, it's smart. I 've attempted to add some new dimensions to what a safari can provide and how to protect yourself. Even when using mosquito netting at night, tickling notwithstanding, it's still necessary to take an anti-malarial drug. Of all the dangers in Africa, one stands out the most: The mosquitos! Malaria is endemic in East Africa. My body can't tolerate mefloquine, the standard preventative drug. I take 100 milligrams of doxycycline daily when on safari, the recommended alternative. The two most important tools to bring along on a safari are a Swiss Army or other multi-tooled knife and a good LED flashlight. Maybe two flashlights!

Mt. Fuji

Chapter Four

Under The Rising Sun
The Atomic Double-Cross and Altered DNA

Prologue

I had three uncles who fought in the Pacific in World War II, and an aunt who served as a combat nurse in the Philippines. All four survived and led healthy postwar lives. My aunt Jeanette, the army nurse, just turned ninety-eight this year (2019); her handwriting is exemplary and she is as lucid and clear-spoken as ever. I learned much about the war from her and my uncles.

This story includes horrific details about the war; Uncles Stan and Don were particularity insistent that I write down what they told me. Italics highlight specific information they had explained. It's *not* something that should be read to children!

The U.S. congress declared war on Japan on December 8, 1941—the day after Pearl Harbor was attacked. For America, this began World War II in the Pacific—ahead were forty-five months of bloody fighting from one island to another. Okinawa was the last to fall on April 1, 1945—resulting in the killing of 12,500 American soldiers, sailors, and marines. The total for the war in the Pacific at that time was over 64,000 Americans killed and over 212,000 wounded or missing. But this was just a fraction of the losses expected for Operation Downfall, the codename for the U.S. land invasion of the Japanese mainland. The staff of U.S. Secretary of War, Henry Simpson, estimated American losses would be between 1.7 to four million casualties, including 400,000 to 800,000 fatalities—plus five to ten million Japanese fatalities!

When President Franklin Roosevelt (FDR) died on April 12, 1945, Vice President Harry S. Truman became president and was confronted with these estimates. FDR had kept Truman largely in the dark, away from critical war planning activities and updates from the Joint Chiefs of Staff.

To worsen matters, Truman was confronted with another reality: At the Yalta Conference in February of 1945, Joseph Stalin, Winston Churchill, and FDR met to discuss how to divide Europe, since they were certain Germany—who they had mutually defeated—would soon surrender. FDR, knowing Simpson's casualty estimates for Operation Downfall, suggested that Stalin move his million-man Soviet army across Russia—no small task—and help the U.S. defeat Japan. Stalin agreed after giving it some deep thought, probably visualizing himself smoking Cuban cigars and drinking Stolichnaya vodka on the tropical beaches of Guam and Saipan the following January.

The idea of Stalin helping to defeat Japan and then taking part in dividing it up after the surrender was unthinkable for Truman—but so was the unimaginable death and suffering of 1.7 to four million Americans. It was clear to Harry that he had to advance the plans to invade Japan before the Soviets joined in; and hope to God for a miracle that would speed the Empire's defeat. Double-crossing Stalin was the least of Truman's worries—FDR had made the deal.

So, Truman had his head in a vise: invading Japan was the only way he knew to end the war, and he couldn't procrastinate. The Japanese had made it clear they would die before suffering the dishonor of surrender. By August of 1945 the Russians would be fighting the Japanese in Manchuria. It was just a hop and a skip to the Japanese mainland from China. Truman's dilemma was quite clear: How quickly could the U.S. defeat Japan before the Soviets tried to join in, and in doing so, not suffer anything close to the losses the Simpson estimates predicted?

Alamogordo, New Mexico, July 16, 1945

Called the Trinity Site, it's located in the New Mexico desert north of

Alamogordo, where a "gadget" invented by Doctor Robert Oppenheimer and his team of scientists will be tested. The time is 5:10 a.m. or "zero minus twenty minutes." Five miles away, scientists in the control dugout are making last-minute checks. They can see a brightly lit tower in the distance; strapped to it is a metallic sphere ten feet in diameter, wrapped tightly with wire. Each scientist has access to a special welder's-glass shield which will allow them to safely observe the test.

Oppenheimer, a basket case even under normal circumstances, is as nervous as a pregnant nun at confession. He gulps black coffee and smokes cigarettes as he fidgets in the control bunker—four miles closer to the tower than the control dugout.

At zero minus two minutes, he grabs his welder's shield, steps out of the bunker and lies face down on the ground. The gadget strapped to the tower is a mile away. A warning flare is ignited.

At zero minus ten seconds, a loud gong sounds a warning: 9, 8, 7, 6, 5, 4, 3, 2, 1, 0 . . .. A flash so bright, at least ten times brighter than the sun (some later claim fifty times brighter) lights up the New Mexico desert. A second flash comes from returning photons that had entered the ground during the initial flash. Looking up, the scientists see a giant fireball through their welder's shields, followed by a tremendous purple mushroom cloud that rises high, pulling tons of sand into it. Instantly, ten million degrees Celsius incinerates the steel tower and desert within a half-mile radius of the explosion. Just over ninety seconds later, an eardrum-busting boom and shock wave crash the air and vibrate everything. Planet Earth has just experienced the detonation of the first atomic bomb—an explosion equal to the force of 21,000 tons of TNT. Later that morning, Truman would hear about it in a telegram from Oppenheimer.

Harry S. Truman now has, if he so chooses, the means to open the vise that compresses his head. But it will be quite controversial to say the least!

The ninety days from mid-April to mid-July 1945—the beginning of the Truman presidency to the A-bomb detonation at Trinity—moved exceptionally fast for Harry. He would later be quoted as saying: "A president either is constantly on top of events or, if he hesitates, events will soon be on top of him. I never felt I could let up for a moment." How true that was. He knew the buck stopped with him.

Two Japanese cities were destroyed and over 100,000 people were killed instantly by two atomic bombs in August of 1945. Another 130,000 were wounded, most of them mortally. The United States dropped the first bomb on Hiroshima on August sixth, the second on Nagasaki three days later. Harry Truman, as commander in chief, authorized the bombings.

A large percentage of the Japanese military fought the war without a scintilla of concern for ethics or human life. They, starved, shot, drowned, burned, beheaded, and disemboweled many thousands of American prisoners and innocent Asian people. Uncle Stan (my namesake) believed a horrific, evil force had awakened in much of the Japanese military. *He told me he knew American prisoners who had witnessed Japanese officers and infantry soldiers being injected with a serum—the prisoners believed it induced incredibly evil behavior.* As you read on, you'll learn that this serum was probably sleeping medication that followed a separate process supposedly transforming the Japanese soldiers.

It's known that the Japanese torture-and-kill philosophy began at the top of their command structure with the generals, admirals, ministers, and Emperor Hirohito himself. Much of it was propagated by the Prime Minister, Hideki Tojo, who refused to honor the Geneva Convention. *A particularly horrific technique involved strapping down a prisoner in the sitting position with his anus directly over a bamboo plant that grew thirty-five inches a day. By noon the bamboo had penetrated well into the prisoner's small intestine and was on its way to the liver and lungs. It took a prisoner a long time to die, while constantly begging to be shot.*

When I was in the Philippines in 2001, I was shown photographs from 1942 of two U.S. marines who had been hung by their tongues until dead. This punishment was for the "offense" of revealing only their name, rank, and serial number. These were the most shockingly gruesome photographs I'd ever seen. My father, as a police detective, took photographs of revolting suicides and murder victims during his thirty years on the Milwaukee Police Force and I got to see many of them.

General Douglas MacArthur, commander of the U.S. Pacific forces, knew such horrific behavior had to be documented. He enlisted the help of translation specialists in ATIS (Allied Translation and Interpretation Service) to decipher 350,000 Japanese files on innocent Filipinos who were burned, maimed, and murdered in various ways after the fall of Manila. *The files proved, with unimaginable, inhumane photographs, that young women were forced to have sex with up to sixty Japanese soldiers a day. They were designated "comfort women." After losing their appeal, they were killed, butchered, and eaten. The Japanese infantry regularly ate U.S. prisoners of war but preferred to eat expired comfort women. Cannibalism was rampant.* After the Japanese unconditional surrender on September 2, 1945, MacArthur used this evidence, complete with its extensive photo gallery, to prove to unbelievers in the Japanese government the horror they had been party to. In Washington, the Pentagon was furious when they found out he did this, but not doing so would have been a historic mistake.

Testimonies of similar horrid events throughout the war certainly affected President Truman's decision to drop the atomic bombs. This all had to end!

Between 1995 and 2014 I travelled to Japan on business and had numerous chances to meet and become friends with many fine Japanese people. Not one of them fit the above characterizations and behavior in any way. Not one.

Surprising to me was their desire to discuss the Hiroshima and Nagasaki bombings. The Japanese were curious how Americans felt fifty years after the fact. They were aware that a Gallup poll taken in September of 1945 had indicated eighty-five percent of Americans approved the bombings. Did the American people still feel that way?

I explained that the American's I've talked to spanned four generations and about half believed the bombings were justified. Baby Boomers were the most supportive, Millennials the least. The youngest Millennials seemed the most indifferent, no feelings one way or the other. I called it the X-Box effect: You never really die in X-Box—you always come back to fight another day.

Boomers understood that two million Japanese soldiers had dug in along the shores of Japan's South Island in the summer of 1945, ready to defeat any invasion. It was clear that the atomic bomb saved the lives of millions of Japanese—estimated at five to ten million—by forcing a fast, unconditional surrender. Japan had already lost a staggering 2.1 million military personnel in the war, plus a million civilians. By July of 1945, Japan's once-mighty navy had been destroyed. Its army's ability to resupply ammunition, ordinance, and fuel was history, and ninety percent of its aircraft had been blown to pieces. It was over for the Land of the Rising Sun—they just didn't want to believe it.

America provided warnings prior to dropping the atomic bombs. Millions of warning leaflets (63 million to be exact) were dropped on Japanese cities by U.S. planes—warning citizens to evacuate, that total destruction was imminent. Possession of a leaflet was cause for arrest in Japan. People were ordered to fight to the death to honor their godman, Emperor Hirohito. In 1945 the government provided the citizenry with long wooden sticks—sharpened at both ends—to be used as weapons against the American invaders. The limited number of guns were just for the soldiers. Japan would never surrender.

It was in 2003 when the friends I had made in Japan started asking a really tough question. They had long understood that Japan had to surrender and, although very hard to swallow, that atomic bombing of a city was necessary to shock Hirohito and his generals into surrendering. But why was it necessary to destroy two cities with atomic bombs? They believed it was wrong to destroy Nagasaki.

I needed to understand more about why we bombed Nagasaki. What went into the decision and why was it chosen as the second city to bomb? Was it reasonable to give Hirohito only three days after Hiroshima to convince him and his general staff to surrender? A lot had been written about the bombing of Hiroshima, not as much about Nagasaki.

Nagasaki, Japan—August 9, 1945

The morning of August 9th was cloudy and cool. Black-tailed gulls, squawking loudly around the docks, were having yesterday's udon noodles for breakfast. Two young schoolboys threw the noodles high into the air as the birds dove and swerved to catch them, competing vigorously with each other. Looking like worms, the long noodles would wrap around a bird's bill when it caught one, hampering its ability to squawk for a few seconds. The boys giggled at this; they often left early for school to allow time to feed the seabirds.

Nagasaki's shipyards and commercial docks were in the southwest part of the city and had escaped much of the U.S. conventional bombing in prior weeks. Preferred targets had been production facilities for ordinance and military equipment. Small buildings and houses were old-style, nineteenth-century timber constructions, some plastered, some not. Nagasaki, unlike Hiroshima, had grown over the years without zoning restrictions, so residential housing was built next to factories.

The seabird-feeding boys were on their way to school when they passed a line of workers waiting for the shift change at the Mitsubishi Steel and Arms Works. The line filled the narrow street in front of the taller boy's house which led to the factory. Jiro was the tallest boy in the elementary school, he had intended to stop and see how his mother was feeling before going on to school; she'd had a migraine headache that morning. But, seeing the line of men blocking his house, he decided to wait until lunchtime—school was nearby.

Each of the boys had a sister who worked at the factory. Their respective uncles were in the Japanese Imperial Army fighting the Soviets in Manchuria. The Japanese 134th Anti-Aircraft Regiment maintained four batteries of guns next to the Mitsubishi factory complex. The boys had wished they were two years older; at fourteen they could push carts of 7-cm shells from the armory over to the soldiers who shot them at enemy planes.

Tinian Island—August 9, 1945

The B-29 Super-fortress, nicknamed *Bockscar,* takes off just before 02:00 carrying an atomic bomb code-named *Fat Man*. The bomb, the size of a Volkswagen, weighs 10,300 pounds and brings the take-off weight of the B-29 to a hazardous sixty-seven tons. The plane, stripped of all armor plating and defensive turret guns, barely accommodates the weight of *Fat Man.*

The full length of the main runway on Tinian Island is required for the B-29 to gain enough ground speed to produce sufficient lift in the humid, tropical air. This island is in the Mariana group north of Guam—six hours flight time to Nagasaki. Major Charles Sweeny pilots the monster Boeing aircraft.

Five B-29s are used for this second mission, same as for the Hiroshima mission. Only one carries an atomic bomb; two fly ahead to report the weather; two fly behind carrying instruments to measure the energy and radiation of the blast and to photograph it.

Three days earlier, *Little Boy,* the first atomic bomb used in warfare, had been dropped on Hiroshima. *Little Boy* and *Fat Man* are significantly different in design—the latter is much more powerful [1].

Fat Man almost detonates prior to takeoff on Tinian Island! Previously armed on the ground, unlike *Little Boy,* which was armed after takeoff, the firing mechanism for *Fat Man* is compromised by improperly inserting an electrical cable after the bomb is loaded into *Bockscar.* The two nuclear engineers who almost sent Tinian to the bottom of the Mariana Trench require several neurotic hours to rewire the connections with zero tolerance for another mistake.

Following the cable-insertion blunder, it's discovered that an on-board fuel-transfer pump is malfunctioning and will not be able to transfer 640 gallons of backup fuel to the B-29's main fuel tank. With no time to replace it or change planes, the B-29 will carry this gasoline as dead weight, roundtrip.

The planned target is Kokura, halfway between Hiroshima and

---

1 Unlike the gun-type atomic bomb dropped on Hiroshima, much more sophistication was required to produce the implosion bomb used on Nagasaki. The bulbous, eleven-by-five foot *Fat Man* bomb contained at its center a 6.19 kilogram, spherical core of plutonium alloyed with gallium. Surrounding the core was a spherical complex of pushers, tampers, an "Urchin" mechanism, and, ultimately, the symmetrical explosive devices that produced the final look of a large soccer ball—a truncated icosahedron.

At detonation the plutonium core is compressed by synchronous, conventional explosions that produce a supercritical-plutonium density exactly timed with the crushing of a polonium-beryllium "Urchin" which injects billions of free neutrons into the core, assuring a chain reaction and nuclear explosion. The big secret was to have all this synchronized such that there would be enough time, in microseconds, to allow chain-reaction fission to occur before the bomb blew itself apart.

Fat Man had more future potential and could deliver larger yields than the *Little Boy* bomb used on Hiroshima; but was far more sensitive to carry, arm, and detonate. Doctor Robert Oppenheimer needed to know that *Fat Man* would function when dropped five miles high from an airplane without a parachute. The disappearance of half of Nagasaki was proof that it could.

Ironically, the successful test at the Trinity Site in New Mexico was with a *Fat-Man*-bomb type that was detonated from a tower on the ground. The *Little-Boy* type was never pretested!

Nagasaki, where munition, arms, and steelworks factories are integrated among historic shrines and tree-lined streets.

Six hours out of Tinian, Major Chuck Sweeney circles *Bockscar* above Kokura for fifty minutes, attempting to rendezvous with the camera-carrying plane which is nowhere in sight. *The Great Artiste,* the instrument carrying B-29, is behind *Bockscar.* The mission operates under radio silence, so nobody is talking. Finally, Sweeney breaks off and heads towards the target site in Kokura.

Kokura is clouded over by weather and black smoke from a steelworks factory purposely burning coal tar. Sweeney makes three passes over the city at 26,000 feet, trying to get his bombardier a clear view of the target site—a large arms factory. Visual sighting has been ordered, no radar.

Fuel is now a factor—it pours into the four 2,200-horse-power Wright Cyclone engines. *Bockscar* is a roaring monster—but for how long?

Antiaircraft fire explodes below them as they search through the clouds and then—suddenly—ten Japanese *Zeros* appear, coming fast from behind. *Bockscar's* gun turrets are empty, no guns, no bullets—too much weight to carry with the five-ton atomic bomb.

Commander Ashworth announces into the crew's headphones, "It's Nagasaki!" as he takes control of *Bockscar,* defaulting to the alternate target. He's in charge of the bomb and his command takes priority. *Fat Man* is alive and armed in the B-29's belly. Ashworth is determined to drop it, somewhere. Speed and altitude are their only defense against the Zeros.

It's monsoon season in southern Japan, and *Bockscar* must fly higher than normal to avoid the weather. How important that 640 gallons of fuel in the reserve tank would be if it could be pumped into the main tank. Four-letter words are voiced by the crew regarding the dead fuel pump. Sweeney, back in control, must push the rpms to outfly the Zeros. He's aware *Bockscar* will not make it back to Tinian after dropping the bomb; his modified plan is to land in Okinawa—except nobody there knows that.

At about 7:50, Japanese time, an air-raid alarm is sounded in Nagasaki—then an "all clear" is given at 8:30. A false alarm; common.

Later, the sighting of two B-29s (*Bockscar* and *The Great Artiste*) at 10:53 does not cause much alarm. An attack by air usually involves two hundred or more B-29s—it took 324 to firebomb Tokyo in March. Two planes must be only for reconnaissance.

At exactly 11:01 a sudden break in the clouds allows Bockscar's bombardier to sight the city's easily recognized racetrack, confirming they were over Nagasaki. He releases *Fat Man* without regard for the planned target site—there was no extra fuel for a second pass. *Bockscar* lunges up from the five tons of lost weight.

"Goggles!" Commander Ashworth yells into everyone's headphones—"the blast will be at least ten times brighter than the sun." Sweeney banks *Bockscar* hard right. All 8,800 horsepower is required to move the B-29 rapidly away from the blast. In forty-three-seconds the bomb detonates at an altitude of approximately 1,650 feet.

The 6.19 kilograms of subcritical plutonium instantaneously compresses into a ball of critical density, about half its original size—the result of sixty-four geometrically positioned, conventional explosions.

In less than a microsecond the resulting nuclear chain-reaction destroys forty-four percent of Nagasaki. Only a small amount of the plutonium fissions and only one gram of it is converted into energy—heat and radiation, according to Einstein's equation, $E = mc^2$. One gram of plutonium destroys a city of 263,000 people.

The bomb explodes above a tennis court, halfway between the Nagasaki Arsenal in the north and Mitsubishi Steel and Arms Works in the south—almost 3 km (1.9 miles) northwest of the planned target. At least 40,000 people are immediately killed. Most are incinerated to elemental carbon, at least 60,000 are critically injured. These estimates are later proven to be significantly higher.

Persons directly below the bomb—the hypocenter—don't feel anything. In less a millionth of a second, they are vaporized. In that near-nothingness of time, their blood boils, their bones and skin melt, and pain-signaling nerve synapses never get to their brains which no longer exist. The atoms that had composed their bodies ionize into gas and get suctioned into the mushroom cloud along with a great tonnage of sand and dirt. Some of it reaches the ionosphere, some reaches outer space.

The *Fat Man* bomb is more potent than the Hiroshima bomb, but its effects are buffered by Nagasaki's proximity to hillsides and the narrow Urakami valley that runs through it. There are 7,500 workers alive at the Mitsubishi munitions factory at 11:01. Forty-three seconds later, 6,200 are dead. Black, shadow-like imprints on the cement floor of the factory reveal the final positions of some of the workers seconds before their incineration.

On the day after the fires subside, a photograph of a tall boy—ninety-percent incinerated, lying prone, holding his throat, naked, and charcoal black except for part of his face—is taken by the Japanese photographer Yosuke Yamahata. The boy's location is near where a school had been close to the Mitsubishi-factory complex. A short time later, such photographs were censored by the American forces—those restrictions were lifted in 1952.

According to a different source—eight weeks after the blast—the total killed by the *Fat Man* bomb was 140,000, fifty-three percent of Nagasaki's pre-bomb population.

The radius of total destruction is approximately one mile, plus another mile caused by the shock waves and fires. The bomb is followed by rainfall containing up to two hundred different radioactive isotopes; it falls on any survivors who didn't get out of the rain.

After surviving the plutonium bomb's five pounding shock waves, Major Sweeney trims *Bockscar's* engines and slowly reduces altitude. The crew don their yellow flotation vests and wonder how cold the ocean will be. They pray for the fuel to last for 550 miles—the distance to Okinawa. The flight engineer estimates their remaining fuel will leave them fifty miles short. Sweeney ignores the estimate; he will land this airplane in Okinawa.

Unknown to the crew, *Bockscar* has been erroneously reported down, and nobody in Okinawa expects their arrival. Since he cannot contact the tower by radio, Major Sweeney brings the B-29 in with all emergency flares burning. Okinawa, under U.S. control for just a few weeks, is a busy place with congested runways. They clear just in time. At first touchdown, Bockscar bounces high, then down hard, as the engines shut down—every ounce of fuel is used.

Tokyo, June 2004

I flew into Tokyo's Narita airport on Delta Airlines, as I have many times, often to connect with a flight to Singapore or Hong Kong—two hubs I used to get to more obscure Asian destinations. This time I would stay in Japan and catch the *Shinkansen* (bullet train) in Tokyo's Japan Station for the hour-long ride to Hakone, which is quite rural. The trains are marvelous: smooth, clean, and fast.

It was a warm, bright June afternoon. Most everything in Japan's big cities is new, fast, and bustling. People walk and talk fast. In Tokyo's Ginza district, it's not uncommon to wait for several traffic light changes before getting to the curb for "your crowd" to cross. But rural Japan makes up for it—beauty, relaxation, and old charm flows through it.

"Doctor Stan! Greetings." Hiro came waving as I detrained. He was a biochemistry professor who referred to anybody with a graduate degree as a doctor. I have a Master of Science degree in biochemistry, so he called me Doctor Stan.

"Kohn-nee-chee-wah, Hiro!" I greeted him, saying konnichiwa fast. Then we bowed to each other twice—a high bow using our neck and shoulders, then a low bow deploying a full bend at the waist. Had I been Japanese, we would have bowed a couple more times.

"Congratulations! You remember how to pronounce hello in Japanese!" Hiro spoke perfect English.

"I practiced in the restroom on the Delta 747. A flight attendant knocked to check if I was okay."

Hiro laughed, then put out his hand to shake mine. "Now for the American way, it requires less exercise."

We shook hands firmly. Hiro Hashimoto was well-framed, five-and-a-half feet tall with silver wisdom-streaks in his onyx-black hair—he loved Ray-Ban glasses that automatically darkened as the light brightened. We'd met in San Diego several years earlier at the American Chemical Society (ACS) convention and became instant friends. I was conscious that I towered over him at six feet tall; he liked to joke that he needed a big biochemist to protect him from the physical chemists. Physical chemistry is a different universe compared to biochemistry; Erlenmeyer flasks are made out of stainless steel instead of Pyrex glass.

"Well, Doctor Stan, are you still okay staying with Hanoka and me at our vacation cottage?" Hiro asked as we drove away from the Hakone station in his Honda SUV.

"Absolutely! I'm looking forward to it." Usually I met with him in Tokyo at the university and stayed at the westernized Marunouchi Hotel.

"You know we sleep on the floor," Hiro reminded me.

"I'm ready, brought an inflatable air matrass that I use when camping out.

"Hah! Hanoka will laugh at that. . .. She will explain how the house works. It will be an adventure for you to live Japanese-style for the weekend."

"A man can never have enough adventures." I smiled.

Mount Fuji's perfect volcanic cone was topped off with June snow; it rose to our left with cobalt-blue Lake Ashinoko in the foreground. I'd been here in the late 1990s when I presented a research paper at a seminar on prawn aquaculture. It was a beautiful disconnect from the rush of Tokyo.

Hakone Town is nestled high in the mountains 106 kilometers southwest of Tokyo, inside the volcanically active Fuji-Hakone-Izu National Park. Natural hot springs proliferate in the wild and within the confines of nearby hotels. Experiencing an *onsen*—bathing in a Japanese hot spring—was something I looked forward to. Hiro had it on our schedule, he and his wife enjoyed sharing a hot mineral water bath with friends.

"Hello, Doctor Stan," Hanoka waved from her garden as we drove up to their charming cedar wood cottage.

"Konnichiwa, Hanoka!" I replied as she came over and handed me a small carton of blackberries. She nodded and smiled as I sampled them.

"What do you think? They're one-hundred-percent organic!" She assured me. I'd spoken to her on the phone when I arrived in Tokyo. She'd alerted me to be ready to perform an organoleptic analysis (taste test) on her blackberries.

"Really special flavor," I acknowledged. "Definitely over forty volatiles: has good balance between bold and subtle aromatic notes, plus just the right sweetness and tartness." She knew I'd studied natural flavors using gas chromatography years ago, and that blackberry was one of the hardest to define. "When cultivated optimally they can contain up to fifty volatile flavor compounds—the molecular formulas of some are still unknown," I explained, then ate a few more. Hiro reached over me and grabbed several for himself.

"Now you eat my berries." She gave him a sly smile. "Only because Doctor Stan said they're good."

"Forty-nine volatiles." Hiro smacked his lips.

"Now he's a gas chromatograph. . .." She shook her finger at him. We all laughed.

Hanoka was a delightful woman who loved gardening. She was retired, but had been active as a botanist working for the U.N. in the 1980s. Her mission was to get various berries growing throughout the Pacific islands.

"Let's go inside, Hanoka will instruct you on the house rules."

"Stan knows about them! Hiro makes me do this with all our guests, even the Japanese."

"Rules are rules." He winked at me. "The latest generation of Japanese boys needs strong lecturing on them."

When Hanoka opened the front door, I knew exactly what to do. I quickly bent down, took my shoes off, and put slippers on—the ones with my initials. I started to put Hiro's much smaller ones on first as a joke. Hanoka laughed and clapped.

I followed her in, walking on tatami floor mats into a large room where the walls were covered with sliding panels. "This is our living room," Hanoka announced. There was nothing in it except two low profile chairs. Glass skylights provided natural light. "These panels are shoji screens, made of translucent rice paper set in wooden frames. They open very easy, be careful not to push or pull them too hard." She demonstrated. One set opened to a large window showing the garden; another opened to various storage shelves hiding a TV, computer, folding table, seat cushions, books, etc. I didn't see any junk!

I followed Hanoka with Hiro behind me into the next room that had a large, square, wooden table just off the floor about a foot high; seat cushions with back support were placed around it. "This is our dining room; it's called the tatami room. We sit on the floor; there is electric heat under the table. We have no central heating; electric heaters keep us warm." She opened a sliding panel that had dishes, napkins, and silverware behind it.

Next were two bedrooms: the larger one was Hiro and Hanoka's—it had a single dresser next to a Buddhist altar with photographs of Hiro's and Hanoka's deceased parents. Tatami mats covered the floor. Futons, pillows, comforters, and a clothes closet were behind shoji screens in both bedrooms. The second one was the guestroom.

"Will you be all right in here?" Hanoka asked me. She slid open the shoji screen and showed me my options. "The electric futon is wonderful, but only run it on low. We don't want the famous biochemist to overheat."

I laughed. "I'll do just fine!" Sleeping outside in a kudu blind in Africa or a hammock in the Amazon were a bit more of a worry.

The bathroom was split into two rooms, one for the toilet (western style with a heated seat!) and one for the bathtub, shower, and sink. Hanoka pointed to a basket of slippers.

"Let me guess," I said. "Change slippers before using the bathroom." Hanoka nodded, smiling. How civilized is that!

"And please note: we share bathwater," Hiro added, probably thinking Hanoka might forget this important economic and environmental policy.

"Let me guess again. First, I shower, soaping up well, washing all parts, then rinse thoroughly before entering the tub." Hiro bowed to me. I never asked what the little yellow bucket was for. Also, the short, wooden stool with a mirror strategically centered on it was interesting.

Finally, we went into the kitchen. It was immaculate like everything else. Ceramic tiles on the walls, rice cooker, electric tea kettle, stove with small broiler (no oven), refrigerator, and cabinets.

"We keep tea hot all day. I bought instant coffee if you prefer," Hanoka mentioned.

"Green tea will do just fine," I assured her.

Then Hiro suggested: "Let's get your luggage and move you in. We've invited another couple to join us for dinner, they are professors too, and I know you will enjoy meeting them. Hanoka and Kira will cook us an authentic Japanese meal."

Hiro introduced me to Hideyoshi and Kira when they arrived; both were professors in Tokyo like Hiro. Hideyoshi taught analytical chemistry; Kira taught molecular genetics and supervised graduate-level genetic research. They had been married for five years and were in their prime.

While Hanoka and Kira cooked dinner, the three of us boys sampled Hiro's favorite sake while sitting on the floor in the tatami room.

Looking directly at Hiro, we held our ceramic cups out slightly, Hiro filled them with the chilled rice wine from a porcelain tokkuri.

"Kampai" he toasted us. Hideyoshi and I followed him a few seconds later.

"Dassai sake: fruity, aromatic and slightly sweet, SMV +2 on sake meter," Hiro grinned. This was the first time I'd had sake served chilled; warm was more common.

"Chilled is best in spring and summer." Hideyoshi took the tokkuri and served Hiro and me. It was my turn next. Rotating servers was a Japanese tradition.

With perfect timing, Kira brought out a scrumptious biodiversity of fresh organic vegetables from Hanoka's garden—lightly boiled and buttered.

I had my chopsticks ready, eyeing the kabocha squash, mountain yams, and the white daikon radishes. It made me recall a sandwich that natives made for me in Malaysia on a fishing outing: thin slices of daikon radishes and red onions on buttered local bread with a touch of salt. Simple and piquant.

The main course was exquisite, it played on the fisherman in me. Each of the ladies carried out a silver platter filled with slices of fresh, seared, bluefin tuna—oxblood-red below the searing. The thought of it still makes me salivate. The tuna was complemented with short-grain white rice, scrumptiously sticky. I had learned to just touch the ends of the chopsticks to the wasabi sauce before latching onto a slice of tuna—only a few milligrams were required to create a perfect, not overpowering, flavor-synergism with the fish.

There is nothing like eating a fabulous dinner on someone's floor to create the right atmosphere for a conversation.

After dinner we all helped with the dish washing and drying, then retired, under starlight, to the living room. Mysteriously, large bean-bag chairs that looked like sea turtles had appeared on the tatami mats—one for each of us. In the center of the room, a small round table with cups and a kettle of green tea appeared. Was this all hiding behind shoji screens during my tour?

Hanoka brought out a tray filled with *mochi* rice cakes—stuffed with naturally sweet *azuki* beans and topped with *kinako* powder—a traditional Japanese dessert.

Hiro poured us tea before starting the conversation. "I'm very happy we can all be together. Doctor Stan, are you comfortable?"

"Like a baby kangaroo in his mother's pouch." I took a bite of *mochi* cake.

"Excellent." Hanoka beamed.

Hiro kept standing while the rest of us partly disappeared into our turtles. "Doctor Stan and I have talked often about the Great War—we've been very explicit and open with each other. That war was the greatest man-caused shock event in human history—over fifty-five million people were killed worldwide.

"Doctor Stan had three uncles and an aunt who fought in the Pacific. They returned home alive. Hanoka's parents were killed in Hiroshima on August 6, 1945 and my parents were killed in Nagasaki three days later. The couples never knew each other. You can see how young they were in the photos on the Buddhist altar in our bedroom. Hideyoshi and Kira lost relatives in the fire-bombing of Tokyo." Hiro's eyeglasses hid his tears.

"Hanoka and I were very young during the war and don't remember much about it. For me it was loud with sirens and explosions and strange smells, like everything was trying to burn. Our house shook frequently. I remember my father carrying me into cold, dark, crowded places—bomb shelters. Now, whenever I smell mold in a cold, damp basement, I remember the war.

"Hanoka and I met in 1970 in Nagasaki. We were attending a

seminar on the lingering effects of the war. It was very ironic. We wouldn't have found each other if not for the atomic bombs.

"Doctor Stan, would you like to say some words? Maybe a brief summary of our conversations over the last three years. Be frank; Hideyoshi and Kira will welcome it." Both Hideyoshi and Kira bowed their heads.

"Sure, Hiro. First, permit me to thank you and Hanoka for inviting me to your lovely vacation cottage and for introducing me to Kira and Hideyoshi," I bowed my head to them. My soul voice went into management mode, ready to offer suggestions.

"I believe Hiro has explained that my career has focused on biotechnology that involves using probiotic microorganisms to advance human health, organic food production, and environmental science around the world. I've been hard at it for over thirty years.

"Studying World War II has been an avocation for me, I have travelled to many of its battlegrounds coincidental to my work. But in no way—knowing how it has affected your families—can I imagine the misery it has caused you all. You have my deepest sympathy."

I paused for a few moments in respect before speaking again. "Hiro and I have discussed whether or not it was necessary to drop a second atomic bomb on your country. This was one of the first non-biochemical discussions we had after we met. It's history that we can't change; Russia defeated Japan in Manchuria on the same day Hiroshima was bombed and was getting ready to attack the Japanese mainland and participate in dividing it up, like they did Europe. Had that happened, Japan would now resemble Armenia rather than the modern Japan it has become. Reducing the time to get Japan to surrender to the U.S. was a major priority, it weighed heavily on President Harry Truman.

"The bomb that destroyed Nagasaki was a plutonium bomb, code-named *Fat Man*. It was much more advanced than the uranium bomb used on Hiroshima, code-named *Little Boy*. A duplicate of *Fat Man* had been tested strapped to a tower in the desert in New Mexico in July 1945—the test was successful! A duplicate of *Little Boy* was never pretested!" I paused for a moment to let that sink in.

"Being more advanced was a handicap for *Fat Man* since it was also more sensitive to vibration and wobble. The Pentagon and the bomb's designer were very concerned that when dropped from an airplane it might fail to detonate. And that would allow Japan to discover many secrets from the parts recovered plus the purified elemental plutonium which they could have attempted to use against us.

However, *Little Boy,* literally a low-tech cannon device, simply fired two chunks of enriched uranium together. No secrets of any significance would be revealed from studying its parts should it not detonate. That's why it was used first. How to enrich Uranium-235—the isotope that causes an atomic chain reaction in a bomb like *Little Boy*—was one big secret that could not be discovered from *Little Boy*'s parts."

The living room was completely silent when I paused the second time. Hiro had heard this from me, the others hadn't. I cleared my throat and continued:

"So the Pentagon's reasoning was this: If they used the untested, low-tech, *Little Boy* bomb and it worked, Japan would have to surrender. Chaos would prevail when they realized that one bomb could destroy a city. And while they were debating how and when to surrender, we would drop the more advanced plutonium bomb, *Fat Man,* from an airplane on another city. Should it not explode, a parachute team would deploy and destroy its secret parts.

"There was also an option discussed at the Pentagon: Dropping *Fat Man* on a deserted island in view of the Japanese mainland, followed by a final ultimatum to Hirohito. It would have satisfied the need to scare the hell out of Hirohito a second time, while testing the more advanced bomb dropped from a plane . . . and it would have saved Nagasaki. The risk, if it didn't detonate, was that Japan might consider Hiroshima's destruction a fluke, and keep fighting.

"For the record, it's my belief we should have given Hirohito and his generals more than three days between the bombings to decide on surrendering. At least one week. That much longer would not have risked an invasion by Stalin's army. And, from what I've studied about the emperor, another four days of nervously pacing the Imperial Gardens would have likely convinced him to surrender."

I then summarized many additional aspects I had learned about the Great War. A dynamic Q and A developed as Hanoka made more tea.

I finished my comments with this: "We can't reverse the unimaginable torture and murder carried out on the thousands of American prisoners and local peoples by many in the Japanese military. It was a shocking, crushing time for human civilization. For me, the most confusing aspect was: How could so much evil exist within the Japanese military? I don't see any of it in the Japanese people of today. None whatsoever! So, where the hell did it come from?" I used a napkin to wipe the sweat off my neck.

Hideyoshi spoke up: "I believe my wife's research has resulted in an unexpected discovery that will shed some light your question. Maybe an answer?"

We all readjusted ourselves in the turtle chairs and drank more tea. I noticed that the Orion constellation, The Great Hunter, was visible through one of the skylights. I needed to go change into the bathroom slippers.

Kira was about thirty-five years old, thin, lightly complexioned, and slightly taller than her husband. Her delicate facial features were dramatically highlighted by short black hair that jutted out evenly around her face, held back by a jeweled headband.

"Thank you for your comments, Stanley. Both Hideyoshi and I are amazed by all the work you've done with probiotic microbes and all the places you've been to in the world. And by your interest in the Great War and how it has affected humanity."

She paused and drank some more tea. We all did.

"My molecular genetics group at the university has discovered some startling evidence that connects evil-aggressive behavior with certain genes. We think it has to do with awakening a special group of sensation-seeking genes that are rarely active. In fact, these genes may exist in our "junk DNA" which is not considered part of the human genome's 20,000 genes." [2]

2 Remember, this dialogue was in 2004 and the human genome project had just concluded; scientists worldwide were just starting to apply its findings. There was frequent confusion over how genetic messaging could be altered, if it was even possible. Claiming that junk DNA (non-coding DNA) could become active in some way was not generally accepted. It was very expensive to get full-genome analyses back then. What I reported above was from my notes made after the meeting in 2004, not from what we know now in 2019. Bottom line: Genetic messages can be altered, and several mechanisms are available.

She paused. I was calmly sitting in my turtle.

"I understand that less than ten percent of our DNA is required for a human being to develop and function, ninety percent or more is considered non-functional junk. Is that about right?" I asked.

"That's roughly correct—at least according to the scientific establishment," Hiro answered.

"Yes," Kira agreed. "The establishment has been the biggest thorn in our butts to get funding for this work. Our twenty-thousand genes only require 1.2 percent of the 3.2 billion base pairs (two nucleotide molecules = one base-pair) that comprise human DNA."

"It's absurd to believe nature would be that inefficient and wasteful!" I interjected.

"Yes, I agree," Hiro declared. Hanoka nodded.

Kira continued: As we all know, humans have forty-six chromosomes, twenty-three from each parent, and each contains a different strand of DNA. Genes are mixed up with junk DNA in the single strands. Better said, gene DNA is mixed with non-coding DNA. Under certain circumstances, the non-coding DNA can become active and exert controlling effects that modify existing genes or it may even substitute as a new coding gene."

"Wow." I had to let that sink in.. . .. I switched back to kneeling on the turtle. I knew that genes directed the production of proteins in our bodies, including all enzymes. And they did this by contacting messenger RNA that then delivered the genetic message—protein production orders—to the cell's ribosomes where proteins are made from amino acids. It's Biochem-101.

"Here's where the establishment has gone wrong," Kira added. "Not all genes code for proteins and non-coding does not necessarily translate to non-functional! We now know that molecular switches exist in certain runs of non-coding DNA—switches that when turned on, can change the function of a particular gene or group of genes," [2].

"That's a double wow," I said, scratching an ear.

"It certainly is," she acknowledged. "And the change can be dramatic. For example, a gene that normally codes for enhanced mental acuity may be altered to code for aggressive behavior."

"So, here's where things get interesting," Hideyoshi pipes in. "Kira will now explain terrorist rats."

"What?" I blurted unconsciously.

"Evil rats: white ones with red eyes." Hiro added some humor. All albino lab rats are white with red eyes.

Kira cleared her throat. "The rat experiments are quite a bombshell. One of my more nonconventional students, Ikki—a ham radio operator and electronics buff—discovered that certain UHF radio frequencies, when pulsed, could activate what he called "warrior genes" in the rats. These genes were either hiding in the junk DNA, or they had been coding for some other function before being altered by the radio waves.

"Rats activated by the pulsed radio waves are living terrors. They ravage their cages, kick food dishes over, paw sawdust out, and bite at the wire latch on the cage door. When taking them out of their cages, I wear welding gloves and use tongs. The rats urinate and try to bite me through the gloves. When placed with normal rats they charge and kill them—biting their necks."

"Do radio-wave-activated rats tolerate each other?" Hanoka joined the conversation, she had been listening intently.

"Strangely they do, up to a point—as long as there are enough normal rats to terrorize. The radio-wave-activated rats enjoy orgies of uncontrolled demonic behavior," Kira explained. "They become truly evil creatures!"

"How many times have you repeated this experiment?" I asked Kira.

"Many—over fifty times!"

"And all show the same results, terrorist rats are always created?" I pushed some of the beans in my turtle's bean bag to the opposite side—for more back support.

"No! Only if we repeat the correct UHF frequencies and pulse rates within certain limits. When Ikki ran the first experiment, he accidentally discovered these the first time, lucky coincidence."

My soul voice spoke up, "There are no coincidences!"

"Doctor Stan, besides being a good biochemist, is a student of the metaphysical," Hiro revealed, while taking his glasses off and cleaning them. In the room's low light, his glasses' dark tinting had cleared.

"It is quite amazing what Ikki discovered." Hideyoshi looked at me. "What do you think?"

"It's Incredible! This might explain how soldiers acquire evil behavior. Do you know anything more? I challenged him. Uncle Stan's words were ringing in my ears.

Hideyoshi looked at Hiro, and Hiro nodded. "Without being too detailed, I can tell you there is good evidence that Hitler's diabolic medical practitioners taught doctors and medics in the Japanese Imperial Army how to activate warrior genes in field officers and selected infantrymen. And American prisoners and local peoples paid an inhuman price as a result."

We all sat in silence. Finally, Hiro said to me: "Doctor Stan, my friend, we have trusted you with this information. You are the only American of those we know that we trust. We do not know if the technique Ikki used on rats matches what the Germans taught the Japanese military doctors. Rats are not humans. Hideyoshi and I interviewed two nonagenarian Japanese doctors, now deceased, who had exposed officers and regulars during the war to 'German Wave Technology.'" Hiro looked at Hideyoshi, "Go ahead and tell the rest of the story."

Hideyoshi looked down for a moment, then up at me. "The doctors did not know any technical details except that the waves were radio waves, and that only a black box the size of a small suitcase was used for treatments. It had a power switch, timer, voltmeter, and power cord exposed. The box itself was made of metal and welded shut. Two soldiers would enter a phone-booth-size chamber containing the black box. They were instructed to keep their arms at their sides and touch nothing. The timer was set for twenty minutes, thirty for large, heavy men. When the power switch was turned on, the timer started and the voltmeter displayed one hundred DC volts. After treatment, later in the day, the soldiers were given an injection of sodium barbital—which put them to sleep for twenty-four hours." Kira and Hanoka looked like they had just come out of a horror movie. I found out later, this was the first time they'd heard their husbands had interviewed Japanese medical doctors from the war.

"What's your plan to handle this 'new' knowledge?" I asked Hiro.

"Let's talk about that tomorrow; Hideyoshi and Kira will be with us. We'll start the day with a cable car ride near Mount Fuji, then some easy hiking followed by lunch and a boat ride on beautiful Lake Ashinoko. And, of course, bathing in a natural outdoor hot spring—an *onsen*, as we call them."

There was much to think about after I got comfortable on the floor on my inflated air mattress that night. My mind was racing; Kira's rat experiments were a game-changer.

It took a good hour for me to fall asleep. Strangely, I remembered Anne Frank's famous quotation before she was taken to die in a German concentration camp at age sixteen: "In spite of everything, I still believe that people are really good at heart." That thought was worth hanging onto.

Postscript

Nuclear bombs are obviously still with us—thousands upon thousands of them. There have been frightful advancements in the last seventy three years since Hiroshima. Explosive power is now measured in megatons; one megaton is equivalent to one million tons of TNT. A twenty-megaton hydrogen bomb today is equivalent to 1,000 twenty-kiloton Nagasaki bombs. Take a few moments to let that sink in. Russia's AN602, a fifty-seven-megaton thermonuclear bomb, could destroy all of Texas if dropped on Austin.

Radio wave and microwave frequencies do alter behavior in animals, and that research continues in many laboratories. Kira's work in 2004 was the first time I'd heard of "warrior genes" that might hide in junk DNA or be an altered form of an existing gene(s)—just waiting to be activated by properly pulsed specific radio wave frequencies. Scary stuff. We know a lot more about "junk DNA" in 2019, and it's no longer anti-establishment science.

Whether the two nonagenarian, Japanese doctors told Hiro and Hideyoshi the truth regarding the black-box device, I don't know. The black box may have been made in Japan, and the story an attempt to pass the blame for Japanese evil behavior onto the Germans. Hideyoshi felt certain that the old doctors believed it produced real effects and that it came out of the Wehrmacht. The really scary thing is the possibility that "German Wave Technology," if it exists, is probably known to most of the G-20 countries by now—and kept under wraps for self-defense when all else fails. You can imagine the headlines in a terrorist, New World Order: "A.I. Robots Attack Radio-Wave Radicalized Humans." Where would the good hide in such a world?

My travels have brought me the joy and adventure I've always hoped for, but during these joyful times, often a dark strangeness would surface: Unplanned, unexpected, unexplained, and largely unwanted—experiences that presented unknown (to me) realities. Confirmation of what is truly real from what is false and twisted, is where I am now, in the last three months of 2019. I have to be ninety percent confident of validity before I take any dark-strangeness experience into one of my published stories. That is the case here.

I'll keep writing and searching for one-hundred percent truth.

One thing is very clear: goodness needs protection! And "wrong" behavior needs to be judged on a scale where many of mankind's more trivial infractions are just ignored. Forget about them! "Sticks and stones *can break your bones, but names will never hurt you." Humankind needs to make some critical adjustments with God speed.*

Plaza Mayor, Madrid, Spain — on a sunny day

Chapter Five

The Rain in Spain
Rebellion—Dancing—Bullfighting

Prologue

In the spring of 1985 as Technical Director of a Minnesota-based animal-health company, I headed a research team that developed health care products for veterinary use; large-animal veterinarians helped us evaluate them. We had developed a probiotic gel packaged in plastic syringes that dispensed probiotics (beneficial bacteria) directly into an animal's mouth. It was formulated for horses, cattle and sheep, with emphasis on their foals, calves, and lambs. Neonates are vulnerable to oral and intestinal infections caused by *E. coli* and other germs. Probiotics prevent these infections by competitively excluding the germs and represent a natural, non-drug approach to treating infections. Veterinarians in Spain were the first to test the probiotic gel in Europe.

Madrid, Spain—April 1985

It was just under a nine-hour flight from New York to Madrid, and Doctor Nicolas Armanaz would be waiting for me at the baggage carousel. After Paris, Madrid was my favorite European capital to visit and conduct business. Taking the escalator down from customs, I could see through the windows that it was raining hard.

"*Buenos tardes, Señor* Randolf," Nicolas' deep Spanish voice rang out.

"Hello, Nicolas," I called out, walking toward him. The bags from my flight had just started coming onto the carousel. We hugged hello.

"It's pouring rain outside." Nicolas pointed to the windows with his folded wet umbrella. "Raining for three days now . . .."

"It was raining in New York when I left—one soggy continent to another." "So, how you say: It is what it is. Farmers in Almería would die for this rain." Almería Province in the south, the setting for many movies, was the driest place in Spain.

"Let's go whistle a taxi!" Nicolas started pulling my suitcase. I knew to just follow him. He was in his forties, short and stout with a full beard and in his prime as a large-animal veterinarian. His veterinary company tested and distributed animal health products in Spain.

"El Hotel Palacio," Nicolas told the taxi driver, glancing at me to be sure. I nodded.

"I know you like that place," Nicolas affirmed.

"Sure do! It's across the street from the Prado museum's five thousand masterpieces."

"Hah, I remember now you were angry at yourself for missing so many art history lectures at the university."

"In undergraduate school, it was a humanities option—the lectures were always scheduled early in the morning. I survived by studying picture books from the Prado, Louvre, and Rijksmuseum—and a girlfriend's lecture notes."

"All the best paintings are on the first floor, like Goya's full-frontal, naked lady—*La Maja desnuda.* The Catholic church refers to her *now* simply as an unclothed woman with no concealments," Nicolas explained, grinning.

"Now" is the key word," I affirmed. "In the past you could get excommunicated for staring at her." Nicolas, a devout Catholic, keeps track of the Vatican's pronouncements and denouncements.

Certain masters had dedicated areas in the Prado, like Ruben, El Greco, Velaquez, Rembrandt, and, of course, Goya. But the Prado's collection was also rich with paintings by Fra Angelico, Raphael, Tintoretto, and Titian. Being an aficionado of art-worthy photographs, it takes a while for paintings to grow on me. Frequent business trips to European capitals helped encourage the process. Unlike science, which is constantly learning and revising itself, paintings are finalized creations.

The taxi's windshield wipers slapped fast and missed spots as the rain continued. Abe, our driver, sang in some language, not Spanish. It didn't sound familiar.

"He's singing in Basque, telling God to stop the rain because it's an aggravation," Nicolas translated.

"I guess that's called for, since the rain won't earn him tips like it would in New York," I admitted. Tipping taxi drivers in Spain was not common practice, rain or shine. I'd been here before.

"You're learning! When in the rain in Spain you pay the same."

I laughed. "In New York, if you need a taxi in the rain you wave a hundred dollar bill to increase the odds of it stopping."

Nicolas laughed, "My Spain is a different world."

A sharp turn onto the Gran Via Boulevard caused Abe to change octaves. We were now on Madrid's equivalent to "Fifth Avenue combined with Broadway," a description made by Ernest Hemingway (Don Ernesto in *España*) who spent close to a decade of his life in Spain. The Gran Via, wide and busy, runs through the center of Madrid.

When we came to the Plaza de Cibeles, we followed the circle around to the road called Paseo Del Prado. The Palace Hotel was up ahead. Its main entrance faced the Neptune Fountain, where the God of the Sea himself, spear in hand—totally wet in the rain—was looking to spear something. It was time for a glass of sherry.

The Palace Hotel, built in 1912 and strategically located in Madrid, had a long history of boarding dignitaries and celebrities.

"Hemingway enjoyed this big place and its short walk to the Prado and, for sure, the hotel's vast selection of aged sherries," I reminded Nicolas. Abe had us stopped behind a tour bus that was unloading at the hotel—he was still singing in Basque.

"Don Ernesto frequented the bar in the hotel's lobby," I commented.

"In his day, the lobby bar was smaller and full of cigar smoke. It's been enlarged and remodeled with air filters," Nicolas pointed out.

"He put the bar's small, stuffy version in his 1926 novel, *The Sun Also Rises*."

"Ah yes, I read it. It's titled *Fiesta* here. Three men fight over a twice-divorced woman who prefers to seduce a young bullfighter. Did you ever go to a *corrida*?"

"Once as a teenager in Mexico City, on vacation with my parents."

"Did you enjoy it?"

"It's hard to remember, we didn't have good seats. It was crowded and difficult to see. My mother hated it."

"It's better in Spain. Your famous author friend, Don Ernesto, described a bullfight best: 'It's like having a ringside seat at a war where nothing is going to happen to you.'"

Abe pulled the taxi up to the main entrance of the hotel. A red-capped bellman in a white suit helped with my suitcase.

"*Uno momento por favor*." I had to give Abe a tip. "For serenading us in Basque," I said to him. He understood and smiled big.

I followed the bellman to the check-in counter with passport in hand. Nicolas went to dry out in the lobby bar.

"Order me an *Oloroso* too, I'll be quick."

I knew the type of sherry he preferred; it was dark, rich, and fortified with extra alcohol. The science of sherry winemaking had been perfected in Spain. By law, the white Palomino grapes had to be grown and processed in Spain's Cadiz province. Basic wine ferments are aged in barrels in a sequence of steps by the *solera* system; then fortified with grape spirit (ethyl alcohol) to increase the alcohol content from twelve percent to as high as twenty-two percent. There are multiple sherry formulations that vary in color from white to shades of maroon— dry to sweet.

In my room, I sprayed on deodorant and changed shirts before heading back down the elevator. Nicolas sat at a small table in the back of the bar reading the morning newspaper and drinking Oloroso sherry. Mine was on the table waiting for me—it was 5:00 p.m.

"No. Dammit, No!" He vented loudly. He hadn't seen me yet in the bar's dim light, I hurried over to him.

"Nicolas, what's happening?" I could tell it was about something he'd read.

"They blew up El Descanso restaurant yesterday morning! Look." He handed me the newspaper. I translated the Spanish in my head: "*Restaurant Blast Kills Seventeen near American Air Base.*"

"Eighty-two more were injured!" Nicolas' face was red with anger. I kept reading.

"It's the ETA guerrilla group—those left-wing bastards!" he barked, getting looks now from other patrons. The bartender quickly came over and spoke to Nicolas in Spanish, calming him somewhat.

I kept reading and doing an approximate translation in my head: Twelve Americans were injured and two were killed. All were from the U.S. air base at Torrejon de Ardoz near Madrid. There were twelve thousand U.S. service personnel stationed at the base. The restaurant was a popular breakfast spot for them.

Nicolas slugged down the remainder of his sherry then looked at me. "The police are investigating—that's a laugh. The ETA has already claimed responsibility. They called and apologized to the victim's families! The detonation went off an hour prematurely, killing more Spanish than the intended American military personnel."

"What terrible news!" I'd missed translating the premature part. Too many times in my travels I've heard such news. It's a potent depression pill that takes the shine off any day.

When Doctor Sebastian Garcia walked into the bar, Nicolas and I were deep into a conversation about the ETA—an acronym for *Euskadi Ta Askatasuna,* aka Basque Homeland and Liberty—a violent separatist group. We were into our third sherry.

"*Uno Fino,*" Sebastian called out to the bartender, ordering a drier, paler sherry. Then he reached over the table and grabbed the hat off Nicolas's head and put it on, only partly covering his thick bush of curly hair. Nicolas rubbed his balding head, grabbed his hat back, then gave Sebastian a sign with his arms that only another veterinarian would understand. I guessed it had to do with repositioning a fetus in a cow's birth canal. Sebastian laughed loudly and gave me a hello hug. It was common for large-animal vets to double as comedians.

Nicolas showed the newspaper to Sebastian, abruptly curbing any more comedy; apparently, he hadn't read it yet. He sat down and scanned the article.

"Explain the ETA to Stan—you grew up around the Basques," Nicolas straightened his repossessed hat, making sure his bald spots were covered.

Sebastian proceeded to enlighten me: "The conflict between the Basques and Spanish has been an issue since medieval times. The Basque people feel squashed between France and Spain and want liberation, self-government. They have representation in the Spanish Parliament, but that's not enough to satisfy their far-left minority.

"Things got serious in 1959 when the ETA formed as a separatist group and started killing people. The ETA, analogous to the IRA in Ireland, preaches revolutionary socialism—their goal initially was to gain independence for the Basque homeland by constant protesting, but then, not satisfied, they quickly morphed into a violent paramilitary organization and began lethal bombing campaigns across Spain. The irony—always there's irony—is that the common Basque citizen wasn't helped one iota in twenty-six years of ETA terrorism, only hurt and further degraded in the eyes of fellow Spaniards. In my fifty years on this planet, I can't cite one example of socialism helping the common man for any length of time." After this explanation, Sebastian had to take a breath and wipe the sweat off his forehead. "Nicolas's hat made my head sweat."

Nicholas remarked, "You need a haircut."

Sebastian ran a hand through his hair, then he turned to me. "I'm very sorry to hear about the bombing at the El Descanso restaurant. I know the place, it's eleven miles from here. I can't imagine how many more of your fellow Americans would have been killed had the bomb gone off an hour later—dozens more would have been eating breakfast!"

Nicolas then pontificated: "How to deal with violent minorities that are exponentially smaller than the majority they claim to represent is a conundrum that humanity needs help solving,"

"Having honest and trustworthy local police is an imperative," I had to add, knowing it was easier said than done. I was thinking about the rampant police corruption in Mexico and elsewhere in South America. Humanity does need help making certain adjustments.

"It's time to change the subject," I proclaimed. "Let's talk about our probiotic gel for farm animals. I understand that both of you distinguished veterinarians have good news to share."

That discussion took all of two hours. Nicolas and Sebastian had success stories to add to my mine. We had provided samples previously and the results were impressive.

"Your probiotic gel completely stops watery-mouth in neonate lambs," Sebastian testified. "It stops the oral *E.coli* infection—a serious problem that results in high mortalities."

"And, it prevents scours (diarrhea) in calves during weaning." Nicolas emphasized. "Weaning calves off mother's milk onto milk replacer is a stressful time for them; digestive and elimination problems are common." This confirmed our results in California. We kept discussing the details.

"Work makes me hungry!" Nicolas realized it was 7:30 p.m. and close to the dinner hour in Spain—8:00 p.m.

"I'll second that," Sebastian agreed, finger-combing his hair.

"I'm buying, and I don't want any argument about it." I stood up and finished my Oloroso. These guys had insisted on paying for meals the last time I was here.

"I suppose you even have a restaurant in mind," Nicolas hypothesized as he drank the last of his sherry.

"I certainly do."

"Well, don't keep us in suspense," Sebastian quipped.

"It's near the Plaza Mayor," I hinted, putting my rain poncho on.

"Ah, you remember from last October, El Sobrino de Botin?" Nicolas stood up and adjusted his hat. "I'm ready."

"*Si, Señor*."

"Your Spanish is improving." Nicolas grinned.

I signed the bar tab and we headed to the taxi queue outside the hotel. It was still raining. God Neptune looked hopeful the rain might improve his odds of spearing something in the fountain. Once in the taxi, it didn't take long to get to the restaurant.

"The rain in Spain is mainly in the plain," Sebastian jested, referring to the song from *My Fair Lady*. He held his umbrella high so he could share it with me. Two completely-white, suckling-size baby pigs were hanging in the window; the restaurant was famous for grilling them in a plethora of variations. We went inside.

As the head waiter seated us on the first floor, I pointed up the stairway. "Hemingway wrote novels and short stories upstairs. He would call a good day of writing a "seven pencil day" when all seven required sharpening."

The restaurant's atmosphere had a unique charm. El Sobrino de Botin opened in 1725 and is claimed to be the oldest restaurant in Madrid. "It was Jake and Brett's final stop in the last scenes of "The Sun Also Rises."

"Good bullfight in that novel!" Sebastian had read it, too.

It took a while for the food to arrive; there never seemed to be any reason to rush anything in Spain. The three of us ordered grilled baby-back ribs immersed in El Sobrino's famous sauce. When washed down with a glass of Manzanilla sherry, much lighter and with less alcohol than Oloroso (15% versus 22%), it delivered us into food heaven. We didn't rush. We were in the most Spanish city in Spain according to Don Ernesto. The meal was Perfecto!

We finished the evening at a nightclub featuring flamenco dancing, just off the Gran Via. It was a down-and-under place in the basement of a large white building—you had to know where to go. Rows of wooden chairs at different levels surrounded a semi-circular stage. It was an intimate setting, the exact opposite of a Las Vegas venue, but something you might find in New Orleans. The lighting was subdued until everyone was seated and had a sangria (included with the admission) in hand. Attendants with flashlights directed us to our chairs.

I had little expectation that anything special was about to happen. I had missed seeing flamenco dance shows on previous trips to Spain but had heard they were good. Generally, I was not thrilled by dancing exhibitions—fishing and hunting exhibitions, yes. Dancing, no. Nevertheless, both of the illustrious veterinarians insisted we do this.

Nicolas had paid extra for seats in the front row. I sat between the two of them and quickly became mesmerized. There were three women in flamenco dresses, *traje de flamenco*: tight-fitting above the waist and long, loose, and ruffled below. Brilliant colors! Accompanying the ladies were four men who immediately started drumming, foot stamping, and hand clapping—just enough of a preview before one of the women took center stage. Wearing a stunning white polka-dotted red dress, the young lady began a marvelous dance routine: swirling with a cerise-red cape while stamping her feet in cadence with the men. Her movements were intense and sensual, the white polka dots on her dress appeared and disappeared as the solid red cape flowed over them. Driven by amazing talent and high energy, her arm movements alone produced a flawless story of body harmony. She owned the stage but never disconnected from the euphony of the other six. Passion flowed from her; sexy yes, but beyond that. It was a passion that comes with doing something perfectly. Everyone could feel it—the place was alive.

Then one of the men joined her, a tall young man in black pants and a white shirt with an unbuttoned black vest. They danced together fast, connected, then disconnected. Perfectly. Stamping and clapping, turning fast then stopping to look at each other.

When the woman stepped back, the young man took off his vest and shirt and danced alone. This was when things got extra fast and loud. I could see that the ladies in the audience were pleased. His energy and coordination were amazing.

When one of the other men began playing a flute in an alluringly mellifluous tempo, a new version of their dance emerged. The young woman rejoined the young man and they slowed and flowed together as if they were floating on a serene river in a supernatural place. Unmistakably seductive, very sensuous— quite astounding! And that was just the first act.

Sebastian gave them two thumbs up, Nicolas threw a kiss, I clapped with vigor, humbled by my ignorance. I would do this again.

So much for Day One. Day Two was a very welcome, dry Saturday; clouds were still around but no rain. Sebastian wanted Nicolas and me to visit his family southwest of Getafe, an hour's drive from Madrid. We could discuss further testing strategy for the probiotic gel en route, then relax in the Spanish countryside at his home. But we had to be back in Madrid no later than 6:00 p.m.—they had a surprise for me. It was quite a surprise. Here's how I *remember it . . ..*

Las Ventas Arena, Madrid

Trumpets play into the sound amplification system as the parade begins. Spirited band music honoring famous *toreros* follows. It's a warm April afternoon.

"A *Corrida de toros* (bullfight) in Spain is highly ritualized," Nicolas explains. He, Sebastian, and I have good seats (*Fina* 1 level, in the sun).

Three *toreros* (matadors) along with their *picadors* (lancers) and *banderilleros* (flag men) march onto the brown clay in the arena's ring, saluting the presiding dignitary in the balcony. Thousands of excited spectators fixate on the proceedings. As I'd told Nicolas, I'd seen a bullfight once before in Mexico City years ago with my parents. We were in the cheap seats—high up where it was hard to see.

But now, as an adult and professional scientist, thoughts of the forthcoming event awaken a state of astonishment inside me—excitement at war with indignation. I've heard good things; I've heard bad things. This being only my second bullfight, I had emotional unknowns. But I know to act impartially so as not insult my veterinary friends who, with some irony, honor the "sport."

"The *torero* with seniority is on the far left," Sebastian points to him—the man is dressed in a spectacular, luminous white suit—a *traje de luces.* In that suit, with his wavy black hair and sideburns, he could have doubled for Elvis Presley. But this torero was about to play in a much different kind of concert.

There were three *matadors* and each would fight two bulls—for a total of six bullfights. The bulls, selectively bred for fighting, were at least four years old and weighed from just over a thousand pounds up to thirteen hundred. They were wild animals and would only fight once.

The parade ends and the arena's ring clears except for the senior *matador* and his team of assistants; the two *picadors* are on horseback. This *matador,* Ivan the Great, was permitted to go first by choice and had picked the most challenging of his two bulls for the first fight.

"Ivan the Great is a famous bullfighter! We are lucky to see him," Nicolas declares. My excitement temporarily gets a leg up over indignation. I force a smile.

Stage one, Tercio de Varas

The trumpet sounds and a huge, sharp horned animal rages into the ring. The crowd roars. The monster bull, black as midnight, stops near the middle of the ring and paws dirt, hating what he sees. The matador signals the banderilleros to thrust their capes in front of the bull while he stands back and watches. This gets my full attention . . . it's happening directly in front of us.

"Ivan is studying the bull's reaction to the *banderilleros*: how the animal moves its head, whether it favors left or right, and where in the ring it wants to fight," Nicolas explains excitedly.

A couple minutes pass as the *banderilleros* take turns aggravating the bull with their capes. The matador keeps watching, studying the action. When the time is right, he steps in front of the *banderilleros* and confronts the bull with his large vermilion-red cape, the *capote*. He threatens the bull with the cape, convincing it to charge. The bull's horns pass within inches of Ivan's right hip while he, with unflappable skill, directs the cape to trail over the animal's head as it passes by— the entirety of the bull just misses Ivan. This is called a *veronica*. The crowd yells *olé*! Two more close passes occur quickly before the bull pauses.

"Very dangerous—first passes!" Sebastian is kneeling on his seat. A *matador* usually keeps farther back on a bull's first passes, permitting him to safely assess the animal.

Holding its head high, the bull sees the matador clearly and attacks again. Ivan jumps away at the last second; the bull's right horn almost gores him. The crowd cheers the close miss.

"That's Ivan's style," Nicolas is on fire with enthusiasm. "The *matador* faces great danger when the bull has full sight and energy. No time to practice anything."

The bull continues to make close passes while Ivan skillfully keeps it attacking the cape, not him. The bull imagines the cape to be a living extension of the *matador*; it is the *matador* in the bull's mind.

"Ivan must bring in the *picadors* on their horses now!" Sebastian blurts loudly. "They will lance the *morrillo* (muscle mound) on back of the bull's neck, weakening the muscles so it's difficult for the bull to look up."

I had read this was a vital step in all bullfights. The bull loses some blood, its first blood, and, if the lancing is done correctly, it's forced to charge with its head down—harder to see the *matador.* The *picador's* horses are blindfolded and padded for protection, but they can still get gored.

I watch the first picador stab his long lance into the bull's *morrillo.* The bull, squirting blood, immediately attacks the horse and tries to lift it up and gore it. This further weakens the bull's neck muscles. Prior to the use of padding in the 1930s, many horses were disemboweled during this maneuver. I'd been told that many of the older spectators missed seeing this; they considered the horses dispensable.

"More lancing!" Nicolas shouts. The bull, fully enraged now, is still holding its head dangerously high—seeing the *matador* too well. The second *picador* comes in and lances the bull again. Now there is more blood, but not lethal amounts.

The bull stands facing the *matador,* considering its dilemma.

Stage two, Tercio de Banderillas

In order to further incite the bull and prepare it for the final stage, the *banderilleros* are brought back into the fight. They come at the bull on foot.

The three banderilleros take turns spiking the bull in its shoulders with barbed sticks, each man has two sticks called *banderillas*—literally "little flags" decorated with colored crepe paper which hang from the bull for the rest of the fight. There's more blood. The *banderilleros* are acting for the *matador* at this point and are always in danger of being gored.

Ivan watches his *banderilleros* perform from outside the ring. The twelve- hundred-pound bull has been weakened but is still very much alive and wanting to fight. It desperately wants to kill Ivan and his accomplices. Nicolas and Sebastian are on the edge of their seats. I watch their facial expressions as they watch the fight—they're intoxicated by the drama. I feel a bit disconnected from the reality of it all, like watching an old western movie and rooting for the Indians.

"The bull has not lost that much blood; a couple liters looks like a lot when spread out on the ground. He will not be happy to see Ivan," Nicolas wipes the sweat off his neck.

Stage three, Tercio de Muerte

I see Ivan reenter the ring alone. He's holding a smaller *muleta* (red cloth) in one hand and a sword in the other. The bull immediately charges him in a series of passes. The *muleta* offers less protection than his *capote* (cape) but is necessary at this stage. It allows him to see better when he attempts to kill the bull.

"He has fifteen minutes to kill the bull!" Sebastian asserts. This time limit starts after the bull makes its first pass at this stage of the fight.

The bull is incited by the moving *muleta,* not its red color—bulls are color blind. Ivan has become a jumping and swerving superman of rhythm and motion. He's dancing with a bull that can kill him, separated from it by only a thin red cloth. In a dance of close passes called *faena,* he allows the bull's sharp horns to come as close to his body as possible. One misstep, one miscalculation, and he can be fatally gored. The crowd knows this, they shout "*Olé*!" after each close pass.

"It's time for him to use the fake sword," Nicolas, standing now, points at the bull.

"Not an actual sword?" I'm confused.

"No, he needs to practice thrusting the lightweight aluminum *estoque* (fake sword). He must test a maneuver that will position him over the bull's head where later, he will thrust the real sword between the shoulder blades directly into the animal's aorta. Each bull is different. When he's ready he'll switch to the heavy steel *estoque de verdad* (real sword). This is the most dangerous time for a *matador*." Nicolas looks nervous.

I believe the fake-sword practice is more for the crowd's entertainment than anything else; it generates more close passes. Ivan switches to the real sword after a few minutes. He allows the bull, significantly weakened now, to charge and pass as he calibrates himself for thrusting the sword. He hears and sees nothing but the bull. Electrified with adrenalin, he holds the sword high and the cape low as the bull charges again. At the precise moment the bull's shoulders are positioned correctly, he pierces the sword straight down into the aorta. It's instant death for the bull. It happens fast. The massive animal plummets to the ground, dead [1].

1 The thrusting of the real sword is called an *estocada.* If the *matado*r fails to produce a "quick and clean death" of the animal, the crowd often protests loudly. This can ruin the whole performance and require the *matador* to cut the bull's spinal cord with a different sword called a *verdugo.* If this fails to kill the animal things get messy. Trumpet warnings are issued as the fifteen minute "kill time limit" is neared. An assistant, a *puntillero,* is ordered in to terminate the bull. The *matador* suffers the dishonor of his failure.

"Bravo! Bravo!" Nicolas and Sebastian cry out while clapping hard. I smile for their sake, and clap that it's over.

The crowd is ecstatic, waving handkerchiefs which call for the attending dignitary to reward Ivan. By Spanish bullfighting standards, it has been a perfect performance by Ivan the Great. The Deputy Mayor of Madrid awards him both of the bull's ears and the tail, the supreme award for a *matador*. Devoted fans carry Ivan around the ring. I watch the field attendants drag the dead bull out. I'm told the bull didn't suffer, and the meat would not be wasted. But my soul voice was stilled with sadness.

Sunday morning arrived with beautiful Mediterranean sunshine. The rain in Spain was gone! Abe, the singing taxi driver, would be happy for that.

I had the day to myself. Nicolas and Sebastian had family matters to attend to. The three of us would be visiting rural veterinarians for five days straight, beginning Monday. I started off with a swim in the hotel's pool. The cool water felt good.

I couldn't get over the dream I'd had the previous night: I was a *matador* fighting a monster bull that had its horns on fire, literally. I didn't want to kill the bull, just put the fire out. It would be a friendly bullfight that I could win—a way of fighting boredom that sometimes comes on in a dream. It was a lucid dream—I knew I was dreaming. But I was puzzled, because I was never bored when awake, always immersed in a multitude of activities. My mind had been tricked. Normally it understood I disallowed boredom in my life.

I switched from doing the crawl to a slow breaststroke to keep from disturbing my recall of the dream: *The bull charged. I quickly pulled the bedsheet off the bed and tried to suffocate the flaming horns with it as the bull passed by me. Close! But the sheet fanned the flames instead, stoking them, getting the horns to glow white hot. I could feel the intense heat radiating from them when the bull came at me again. It wanted to kill me; it was not bred to have a friendly fight. Standing up on the bed, half awake, I held the sheet out facing the bull, like a matador would do with his cape. The bull charged again . . ..*

I had the pool to myself and stayed in it for almost an hour. My mind kept going over the dream. What sense did it make? What was my subconscious trying to tell me . . . what boredom could do if I ever let it take control? The real bullfight had served as a catalyst. That, I was sure of. When I could feel the pool's hypochlorite oxidizing my eyes, I knew it was time to get out and locate something to eat—like breakfast.

After consuming a customized three-egg omelet and fresh-squeezed orange juice from red oranges grown in Seville, I was ready to hit the day in the teeth.

I put a new roll of high speed film in my Nikon N-70 and began the short walk to the Prado. There were a number of masterpieces I needed to revisit.

Postscript

The sunshine continued up until my last day, when rain came back with a vengeance. It was blamed on ash from a volcano in the Canary Islands seeding clouds over Madrid. Nevertheless it had been a good week, with lots of horses, cows, sheep, and their offspring happily consuming the probiotic gel. We used ground sweet corn as a flavor in the gel to disguise the medicinal Lactobacillus taste—they loved it.

I came back that September to follow up with Nicolas and Sebastian. They were "over the moon" from the positive results they were getting with the probiotic gel. Once again, it rained the last day of that trip so a second bullfight they had planned for me had to be cancelled. I was happy about that. So I suggested to them that we take in another flamenco-dance show—as my treat! I'd been saved by the rain in Spain.

Yellow Mountain, China —Anhui Province
Flying Rock at Yellow Mountain

Chapter Six

Shanghai Secrets
Progress and Conflict in China

Prologue

On my first trip to Shanghai in 1996, the division of the city between old and new, rich and poor, technically competent and not, was well underway. It had been eighteen years since Deng Xiaoping opened China to western business, economic flexibility, and tourism—contrasting sharply with Mao Zedong's past policies of closed doors, strict controls, and unwavering state socialism. Challenging these evolving divisions was the age-old dilemma of how to control and allocate rapidly expanding wealth without destroying the process that created it. It was a fine edge to balance on . . ..

Shanghai, the largest city in China, seemed to be the most successful at allowing its younger generations develop a successful middle class—self-sustaining, and then some. A class whose success, if not overregulated, would help support the needs of China's general population.

For eight years I traveled frequently to China, stopping in Shanghai and environs on most trips. I worked as an environmental consultant in water pollution control and as a probiotic specialist.

I learned both disturbing and fascinating things about China on these trips—the following story touches on a number of those learning experiences.

Shanghai, June 2003

A breathtaking dichotomy was occurring far below me in the center of new Shanghai as I brushed my teeth while looking out the window of the Hyatt on the fifty-sixth floor. I could see the obvious divide that had developed in the city. To my left across the river were old wooden buildings on cobblestone streets dominated by rickshaws, bicycles, and pedestrians in grey. To my right, stunning sky-piercing constructions of glass and steel sparkled in the morning sun on roadways dominated by expensive cars and well-dressed, fashionable people.

"Greetings, Mr. Stanley!" Terry Ho's deep voice rang out in the lobby when I came out of the elevator.

"Hello, Mr. Ho!" I replied as we shook hands vigorously. I consulted for his company, Ho Environmental, Ltd., dedicated to water pollution control utilizing enzymes, probiotic bacteria, and unique floating aerators.

"Let's go, we can talk in the car." Mr. Ho was not a fan of talking in hotel lobbies. He grinned big and shook my hand again as we took the elevator down to street level. Terry was in his early fifties—husky and smart with strong hands, a square chin, and charcoal black hair. Born and raised in Guangzhou (old Canton) in southern China, he had a degree in chemical engineering.

We went outside and got into his Mercedes. Chang, his driver, had the doors open, waiting for us. Terry gave him instructions in Mandarin's Shanghai dialect.

"Does Chang understand Beijing dialect?" I asked Terry.

"Not so much; there's a significant difference. The Shanghai dialect is actually the original and most common version of Mandarin. All together over two hundred different dialects are spoken in China."

"The written language, the characters, are the same everywhere, right?"

"Correct; it's what holds the country together. An educated person knows about eight thousand characters. You can read the newspaper knowing three thousand or so. *All* of them need to be memorized, each one. There's no alphabet."

"It still amazes me. How many characters are there?"

"Around fifty-thousand," Terry answered while tapping Chang on the shoulder, pointing to something.

"That's mind boggling!" I mumbled.

Chang made a careful U-turn at the next intersection and headed toward the river—responding to Terry's tap. The main driving rule in China was not to hit anybody in front of you—you were automatically at fault if you did. It didn't matter whether a vehicle pulled out in front of you, cut you off, or didn't stop at a red light. Hit it and you are guilty!

Chang drove onto one of the bridges that spanned the Huangpu River, also known as the Whangpoo; both names were appropriate given its malodors. The river divided Shanghai into two regions: the *Pudong* (East Bank—new area) and the *Puxi* (West Bank—old area). Chang pulled over so we could get out and join other river gazers on the bridge's walkway.

"This river needs our help," Terry said. "Dead pigs, chickens, and sometimes people, float by. This upsets the tourists."

"You're kidding," I replied, gazing behind me at one of the world's most dramatic skylines. What a contrast!

"It happens when slaughterhouses upstream get flooded—animals and people drown; untreated excrement further exacerbates the problem." Terry spent two years in San Francisco in the early 1990s and had mastered American English.

The river's brown water circulated below us. Activity from vessels large and small moved the water: oil tankers, tugboats pulling barges, flatboats, junks, ferries . . .. A pink chemical solution squirted from the hull of one of the tankers—probably a degreasing wash. Chemical pollution!

"What do you think?" Terry asked me.

"The river's microbiology is totally out of balance. Gram-negative bacteria are dominant—they cause bulking of the polluting organics preventing them from settling. A rich culture of protozoa is required to get settling, which depends on having gram-positive bacteria, their preferred food. The protozoa feed on the gram-positives, get fat, die, and settle to the bottom. It's a bioremediation cycle that needs to be encouraged," I explained.

"Exactly what we heard at the pollution control conference in Tokyo last December," Terry added. "The Shanghai government is determined to solve this problem; tourism is at stake. They will pay for our help."

"It will be a challenge; rivers are a different breed of cat compared to ponds and small lakes. A place to start is to inoculate and effectively aerate any waste treatment ponds that drain into the river. The inoculation requires trillions of gram-positive bacillus bacteria that must be cultured onsite," I emphasized.

"We'll meet with city engineers on this trip; you'll have your chance to explain Microbiology-101 to them."

We walked the bridge for some distance, continuing our discussion in preparation for the next day. Chang followed along slowly with the Mercedes paralleling the curb, carefully watching Terry for hand signals. Chang was half a foot shorter than me, probably in his mid-fifties, wore glasses, and knew a few English words. He was totally dedicated to Mr. Ho, who treated him well. When Terry gave the signal to stop, we got in and Chang made a U-turn and headed back to the Pudong District, modern Shanghai.

"I'm going to treat you to a special perspective," Terry told me. "And we don't have to go far."

I just smiled as I stretched my neck to see Shanghai's unique sky-scraping architecture; Terry loved to add "special perspective" to my visits. Shortly, Chang pulled into the parking lot of the 1,535-foot Oriental Pearl Tower. This impressive construction, one and a half times the height of the Eiffel tower, is supported by three massive columns that are anchored underground. Eleven spheres attached to the columns house TV and radio stations, a restaurant, executive accommodations, a shopping center, and fifteen observation decks. Brilliant night lighting made it easy for astronauts to identify.

Terry had tickets for the highest of the observation levels—the Space Module at eleven hundred feet. We queued up for the elevator.

Outside the Space Module, we stood on the deck's clear glass floor. Looking down beneath our shoes; we studied the city of twenty-two million people.

"Special perspective!" Terry proclaimed. "It's almost *three times* the size of New York City."

Chang grinned and explained something in Mandarin, pointing east.

Terry translated Chang's comment: "The breeze is from the ocean; the air will be clean today." This got glaring stares from other viewers—I presumed they were wondering how a short, phantom-grey Chinese man, defining common, could be instructing two businessmen in suits in this expensive place.

Before we went back inside to take the elevator down, Terry explained to these other viewers: "Mr. Chang is my chauffeur and confidant. I want him to experience this view of his birth city. Attempt to understand that . . .."

The view had enlivened Chang. He thanked Mr. Ho with a hands-together bow when we exited the elevator. Then it was back into the Mercedes, en route to more special perspective. First days with Terry were always invigorating—mostly play with only a sampling of work. As Chang drove on following Terry's instructions, I asked Terry about a sensitive topic —one he had touched on briefly when we were together in Tokyo.

"Terry, how's your brother doing?" I put my hand on his shoulder.

"He's still assembling Christmas lights," Terry answered, disgustedly.

"In the same prison?"

"Same one: White Cloud Detention Center—Guangzhou. He still sleeps on the concrete floor with fifty other prisoners, sharing one toilet—one hole in the floor."

"Do they still work twelve-hour shifts?"

"Yes, but they get extra rations if they work fifteen hours."

"Any improvement in the food?" I knew this was a big concern of his.

"Still rice fried in pork fat. On a good day they also get moldy turnips." Terry replied, shaking his head.

"Is your lawyer making any progress on getting a hearing?"

"No! I'm going to change lawyers. My brother hasn't been charged with a crime yet; he's just being detained—ten months so far!"

"Do they still whip him?"

"Only if he doesn't make his quota—he needs to assemble six thousand Christmas lights a day. A guard watches him and shouts *Kuai-dian* (faster) constantly—always ready to speed things up by whipping him with a cord containing broken Christmas bulbs." Terry was holding back his anger.

"Unbelievable!" That's horrifying."

Terry's younger brother had been caught having sex with a Taiwanese girl. The Chinese Ministry of State Security (MSS)—the Chinese KGB—had been following the two of them in Guangzhou last September. His brother got detained and the girl was never heard from again. It was alleged that she'd had political connections in Taipei.

"Terry, your brother's situation parallels a similar incident that occurred recently in Beijing where an expat from the U.S. got detained for ordering a massage in his hotel room. The masseuse was a prostitute working for the MSS, so he was detained for having an illegal sexual encounter with a Chinese citizen. A setup! The MSS armed with assault rifles barged into his hotel room shortly after the woman had stripped naked—they took photos! He's still detained as far as I know. My agent in Hong Kong keeps me informed."

"Being detained in China is no day at the beach!" Terry was scarlet with emotion. Keep me informed on the expat."

"I will, definitely."

Looking out the window on this fine June day, I was sorry for Terry's brother. Even Chang, understanding some of what we said, was redfaced. The breezes coming off the East China Sea were fresh and—as Chang had observed— they cleared the city-haze created by the activities of Shanghai's millions. Even on bad days the haze was nothing compared to Beijing's unbearable smog. The politics were different too. Shanghai was more westernized than Beijing, free enterprise more apparent and many of the communist doctrines ignored. This opposed the fact that the Chinese Communist Party (CPC) was founded in Shanghai in 1921. Fortunately, banners promoting Mao

Zedong's "invincible thought" are largely lost relics of that demonic era.

"Ready for more special perspective?" Terry suggested.

"Sure, surprise me." We both needed several kilojoules of happy energy after discussing his brother.

Terry grinned at my quick response; then told Chang where to go. We headed northeast through parts of old Shanghai to the Zhuanghuabang River in the Jing'an District to an alluring temple complex.

"The Jade Buddha Temple!" Terry enthusiastically announced: "Home to the best Buddhist sculptures, including the pulchritudinous sitting and reclining jade Buddhas. These were brought from Burma during the Ching Dynasty in the 1880s. Both have been expertly carved from single pieces of solid white jade."

"You get five stars for pronouncing pulchritudinous correctly," I told him.

"A Buddhist monk fluent in Latin told me it meant 'exquisitely beautiful'—it always gets attention when I use it. Everyone wonders how a Cantonese businessman can know such an English word."

Laughing, I told him he was no ordinary Cantonese businessman. The time he'd spent in San Francisco thoroughly Americanized him.

We stood outside and examined the architecture of the temple while Chang parked the car. Upturned metallic eaves—classic Song Dynasty reproductions—emerged from the roof tops and corners of the three main halls that were connected by two courtyards. The buildings were painted in soothing pastels—very Chinese.

When Chang returned, the three of us entered the Heavenly King Hall with its four Heavenly Vajras (Kings), each five meters high. The first one guarded Buddhism, the second promoted merit, and the other two preached the benefits of far-sight and virtue respectively. None seemed interested in threatening humanity to damnation for trivial sins and misdemeanors!

In the Grand Hall "Nine dragons pouring water" spiraled out of the ceiling, while three golden Buddhas—each four meters high—towered over us. Terry explained the significance of each and how they connected with Buddhism's past which originally was a branch of Hinduism. The Grand Hall was also filled with statues of Buddha's disciples and other guardians of Buddhism—all male, all bald, and all draped in golden garbs—as far as I could tell.

Outside in one of the courtyards, stone carvings of lions and dragons confronted us as we headed to the Hall of Wisdom. Strangely, we didn't see many other tourists. Terry said there were one hundred sixty Buddhist monks somewhere, probably meditating.

When I first saw The Sitting Jade Buddha, I was struck by its beauty and polished perfection. It wasn't a gold-plated statue but a two-meter, single piece of white jade carved such that it could have doubled for a living person. Carving semiprecious jadeite—jade with a Mhos scale hardness of seven—is much more difficult to carve than marble with a hardness of three to five. One mistake and the jade would have been ruined. This is one of the most precious statues of the Buddha that I've seen.

Sitting naked, except for a narrow gold belt inlaid with jewels that wrapped around vital areas, it was perfect. This Buddha's skin tone was light beige from its waist up, to a darker brown for the legs and feet—these were the colors of the "white jade" used and, very likely, selected for its dual-racial character. But what amazed me most was the perfect face: feminine eyes highlighted by glistening eyebrows, invitingly gazing down in contemplation, while a Mona Lisa smile signified perfect contentment. However, the body was more masculine with large hands and feet and was lacking female breasts and a typical feminine physique. The differences were subtle but unequivocal to me, which conflicted with everything I'd been taught about Buddhism—that only males can be Buddhas. Terry was impressed that I recognized this.

"There are many sects in Buddhism; the artist that carved this Buddha in Burma—Myanmar—obviously belonged to one that believed in showing the dual-gender nature of the Buddha's oneness, just as its changing colors depicted a mixed-racial nature. Brought from Burma in 1882, this masterpiece commemorated Buddha's life at the Shanghai Buddhist Association's 1956 celebration of the *2500th* anniversary of Buddha's enlightenment," Terry explained.

"It survived Mao's cultural revolution!" I remarked.

"Yes, where many of the lesser temples and sculptures did not. Buddha's philosophy can be summarized in one of his many quotations: 'All moral sayings and deeds benefiting human life are in concordance with the Dharma.'"

On this trip I was learning more about what motivated Mr. Terry Ho. He's a serious student of Mahayana Buddhism—not as a pious believer but as an investigator seeking truth. He's captivated by the assertion that Mahayana Buddhists follow a path that allows them to achieve enlightenment—the state of Nirvana—in a single lifetime. And they're not plagued with having to endure multiple reincarnations to achieve enlightenment, as believed in Theraveda Buddhism.

Mahayana Buddhism emphasizes that altruistic motives must dominate individual behavior, completely without ego, to achieve entry into the divine spiritual realm of Nirvana. Such an individual is called a bodhisattva: one who willingly delays entering full Nirvana while helping others find their path to it.

The war between good and evil, billions of years old seems to depart from the expectations of Nirvana's Elysian vision. So, if Nirvana—Heaven—is, indeed, a true place, it appears that humans must exist presently in an ephemeral reality where conflict and war are illusions meant to educate our souls as they, hopefully, make their way to Nirvana. Mahayana Buddhists have learned to disconnect from the false reality of these illusions—akin to dispensing with bad dreams that never get resolved.

I knew this from previous discussions with Terry on other trips. He was attempting to confirm whether the Mahayana Buddhists could truly earn a nonstop, first-class ticket to Nirvana and rejoice in Elysian existence forever.

It was time to eat something for lunch. Terry had Chang take us to the Lao Fandian restaurant in old Shanghai—in business since 1875 on Fuyou Lu Street near the Yu Garden. It occupied a three-story house, had several uncrowded dining areas, each with its own kitchen, and a pleasing Shanghainese atmosphere. There was no cacophonous clamor like at many Chinese restaurants.

"They're known for their Bengbang Cuisine. Translation: local food specialties," Terry proclaimed.

"The aromas are enticing. I'm starved," I responded. A well-dressed waiter had seated us on the first floor at a small table next to a high wall covered with impressionist-style paintings—mostly flowers.

Terry apologized for the late lunch. I told him it was my fault for only having tea this morning. Ernest Hemingway claimed no day was longer than one without breakfast.

"What's good here?" The menu was in Chinese characters—some of which appeared to have swirls simulating steam.

"The house special is *ba bao ya,* he said. "You should try it."

"More detail please." I grinned, thinking about the surprise he'd pulled on me at a dinner in Guangzhou last year.

"*Ba bao ya* is boneless duck with eight foods inside, they stuff the duck with diced pork, chestnuts, and ginkgo seeds, then steam it for five hours, twice. It's served on a plate surrounded by rice, beans, and gravy."

"What are you having?" I asked him.

"The *ba bao ya,* of course! With a glass of French Bordeaux. How about you?"

Terry was playing me. He knew I remembered the putrid, fermented *shu tofu* he insisted I try in Guangzhou. Its aroma was exactly that of stale urine mixed with dirty athletic socks. Noxious fumes surrounded the vile 'delicacy' . . .. "It tastes better than it smells," were Terry's famous words. Wrong! It taught me to never underestimate the potential for unpalatable surprises in China.

"I'll have the same, the duck, except your description seems to be short two foods," I replied. I only heard six."

"You forgot the white and dark duck meat itself." Terry smirked.

"Bingo. You got me," Mr. Ho! "I guess I'll have to trust you this time."

"You won't be sorry. The *shu tofu* in Canton was a test of how serious you were about doing business in China."

"I laughed. "Did I pass?"

"Your here right now?"

"True. Minus the *shu tofu!* I'm not sure I could try it again. It was worse than eating monkey brains in Harbin!"

Terry laughed out loud and patted me on the back.

The duck was fabulous. The wine a bit too red for me, but okay; it had travelled a long way.

I slept well in the Hyatt that first night and ate a large bowl of noodle soup and two soft boiled eggs for breakfast—primed for a long day. It would be a travel day at the start: Chang would drive Terry and me two hundred twenty kilometers south to Hangchou on West Lake for business—a three-hour drive. We'd have time to talk.

I found Terry's devotion to studying Mahayana Buddhism alluring; it got me thinking more about Mao et al. I asked Terry: "How did Buddhism survive the Cultural Revolution?"

"With difficulty," he lamented. "I was a young man during the revolution, from 1966 to 1976. Like I mentioned before, some of the temples survived, like the Jade Buddha Temple. Many others were badly damaged."

"By the Chinese military?"

"No—by Mao's Red Guard!—Young men and women in white shirts and red shoulder belts who drove around in pickup trucks flying red flags. They killed the anti-communists: the *bhikkhu*, the priests, the professors, successful businessmen . . . no quarter was given. Corpses were hung on bridges and in public parks. Fear engulfed everyone," Terry explained with emotion.

"Pure Evil! Its details were edited out of the history books! How was the Red Guard organized?" I asked him.

"There was no central organization, so breakaway factions developed and fought each other for territory. One night in Guangzhou, in 1967, two-hundred presumed anti-communists were killed and hung from bridges around the city."

"What motivated them? Ideology? Money?"

"Ideology! Mao's thoughts! Any departure from true Marxist-Leninist social theory was forbidden. Even a half-step back from full communism into socialism was considered anti-communist. The Red Guard would stop people on the street and ask them to recite Mao's writings, like 'Challenge selfishness and criticize revisionism.' or 'Always serve the people.' Very dangerous times—millions died!"

"Was it that bad in Shanghai?"

"Not as bad as in central China. Shanghai, as you know, is the biggest city in China; there are many places to hide. Locals organized and made weapons out of railings and steel fences. Then they ambushed Red Guard trucks, killing the young terrorists while posing to be part of a breakaway Red Guard group. This seriously fertilized a war within the Red Guard—thankfully decreasing some of its effectiveness."

I looked at Terry and didn't say anything more, just shook my head. This was never discussed in the World History I took in undergraduate school.

Hangzhou City

The countryside en route to Hangzhou and its famous West Lake was dominated by rectangular apartment buildings and high-rises, many under construction. Open country was sparse, and shanty towns were not apparent except in the distance, well off the highway. So, in the two hundred twenty kilometers between Shanghai's twenty-two million citizens and Hangchou's eight million, I estimated there were several million more. Lots of people!

Hangchou had taken advantage of its location on the Yangtze River Delta, like Shanghai, to bolster its development and become one of China's most prosperous cities. The road was good, and Chang drove carefully keeping a big eye out for anything or anybody in front of us.

Mountains and tea plantations appeared on our approach to Hangchou, as did more pleasant countryside sprinkled with pagoda style architecture. Hangchou, an ancient walled city, dated back to before the Tang dynasty in 600 AD.

There are thirty-six "West Lakes" in China, but the most tourist-significant is in Hangchou. The scenery around West Lake was impressive—purportedly unchanged from how it appeared in twelfth-century China. Two hundred years ago Emperor Qianlong marked each of the lake's "ten most enchanting scenes" with four-character inscriptions. The scenes included a willow-tree grove populated with singing orioles, a sunset glow over a tall Pagoda, a pond full of hungry goldfish, and lotuses in the breeze—at a place on the lake called the Crooked Courtyard.

Our business focus was to visit these popular tourist sites around West Lake along with four of Hangchou's city engineers and provide guidance on how to control water pollution using beneficial microbes and aeration. The day was still young.

Six of us—Terry, myself, and the four city engineers—joined several Chinese tourists on a boat tour. The boat had a classical Chinese roof with turned-up corners and two guardian-dragon carvings for protection. Terry sat beside me and translated.

Our first stop was by a group of tourists feeding hundreds of koi mixed with goldfish off the end of a dock extending into the lake. Originally dominated by regular goldfish (crucian carp), koi now prevailed; compared to the common carp, they had longer barbels on their chins and were multicolored and valuable.

A full spectrum of colorful koi splashed as the tourists threw feed pellets at them. We watched the feeding frenzy from the boat; the water was solid with fish.

"They're always hungry," Terry said. "The engineers are discussing the frequent feeding that's required to please tourists—often six to eight groups daily. They're worried about the effects of overfeeding. What do you think?"

"Overfeeding is a problem—it can cause bacterial fin rot." I could see several large koi close to the boat that had damaged pectoral fins, signaling fin rot. I pointed this out to the engineers. The water was turbid and polluted below the fish; the lake was calm; there was no aeration.

Terry translated my comments, the engineers asked for a recommendation; they were mechanical engineers, not biologists.

"First thing: Put a floating aerator off the end of the dock where the fish congregate, run it whenever they're not being fed, day and night.

"Second thing: Change the feeding procedure. Big hotels in Japan feed their koi only once daily for no more than five minutes; they use high quality feed containing over forty percent crude protein. I recommend you follow the Japanese: Feed for only one minute each time a tourist group visits."

Terry translated. The engineers talked with him; the tallest engineer thanked me in English. "Tank ou Mr. Stan-lay." Terry and I smiled.

For the next three hours we got a five-star tour of West Lake's "ten famous scenes," finishing at the ubiquitous, subtropical lotus plants with their flamingo-pink blossoms undulating in the day's breeze. While the tourists in our boat took photographs, I explained to the engineers how Terry's solar-powered aerators would oxygenate the lake's water and stimulate beneficial aerobic microbes (aquatic probiotics), thereby effectively reducing pollution and making the tourists happy.

The following day we would study and advise on pollution issues involving the Grand Canal, the largest manmade waterway in China—larger than either the Panama or Suez Canal. The Grand Canal links Hangchou and other cities on the Yangtze Delta with Beijing.

Our hotel in Hangchou was a three-star; the four stars had been booked before we scheduled the trip. However, compared to many other three-star hotels in China, I gave this one a passing grade for attempting to compensate for its shortcomings. The pillow was soft and had been sterilized. And even though the cold water from the small sink was brown, there was a working toilet! As far as bathing goes, they provided a bucket of clean water with a bar of soap, a sponge, and a big towel. But there was no hot water from 7:00 p.m. to 7:00 a.m., and only one dim pole lamp that flickered. Thankfully I could still read because I always packed a spelunker's head lamp. Overall Grade: D+.

It was time to freshen up before the evening meal.

Dinner was a gala event in a large restaurant in downtown Hangchou with lots of patrons and clamor—starkly juxtaposed to last night's more sedate scene. The city mayor and several of his staff had joined the six of us for a Cantonese-style feast, which he knew would please Mr. Ho. The mayor had been told by the engineers that our day on West Lake had been very productive and that Terry and I had advised them well.

All ten of us were seated together at a large, round table containing three lazy Susans that servers were filling with food. Being well schooled in Chinese cuisine from a dozen trips, I was able to identify about half the dishes. The mixture of aromas was always interesting.

"Gum Pay!" The mayor said as he raised his wine glass in a toast. The nine of us followed his lead, raised our glasses, said "Gum Pay" loudly, and drank. The wine, sweet and fruity, went down easily. I was always cautious when drinking an unknown liquid in China. Two of the mayor's three staff members were women, and both spoke some English. The older of the two had long black hair and penetrating, dark eyes. She winked at me after I had swallowed the wine and asked if I would have preferred Chinese whiskey.

"No, the wine is great!" I said quickly, exaggerating a bit. Chinese whisky is one-hundred-fifty-proof ethyl alcohol spiked with acetone. It's poison. I'd learned that in Qingdao a few years ago. "Whiskey is too strong for me." I winked back at her, making sure I didn't specify Chinese whisky—political correctness.

Everyone began intercepting food off the revolving lazy Susans with plastic chopsticks. Most piled it on top of their bowl of white rice before shoveling it down. A waiter had given me a fork but I was chopstick-competent; the real test was whether I could snatch a slippery boiled peanut off a moving lazy Susan with one snap of the chopsticks. I passed the test.

"Raw jellyfish tentacles," Terry whispered, smiling, as they passed by on the lazy Susan. "Dip them in the red sauce."

"I'll mix some in the fish-head soup later," I replied, joking of course. "The stir-fried shrimp are quite good."

"Pleased to hear that," Terry remarked as he chomped down on a grey sea slug dipped in soy sauce. It went like this between us, with me mostly prioritizing "normal foods" like shrimp and mussels. The array of Cantonese foods posed some significant challenges.

The mayor proposed another toast. After drinking the wine down, I saluted the two women with my empty glass. Seconds later it got refilled by the restaurant's wine server, a lady in an elegant gown who I hadn't noticed. There was no such thing as sipping a drink in China, a three-ounce glass (100 ml) provided enough for two toasts. Before heading to China for business, it's a good idea to practice slugging down fifty milliliters of your favorite wine without choking.

The last course was a giant boiled carp—one that had been shown to us alive in a sack before boiling—proof of freshness. It was served whole with its head and tail on. Eyes the size of quarters stared at us. It looked like a monster koi, only mud-brown. I sampled a small amount. Eating dinner in China can be a major adventure!

Terry had reserved a room for Chang and always made sure he had enough cash— the one-hundred yuan pink notes with Mao's mugshot on them—to rent a room or buy a meal. When Terry and I were alone, Chang usually ate with us. In a pinch, though there were rules against it, Chang loved to sleep in the back seat of Terry's Mercedes after snacking on imported Ritz crackers. Some rules were meant to be broken—you just had to know which ones.

The three of us sat in the hotel lobby on our last morning in Hangzhou. We sipped boiling-hot green tea from cups with no handles, mostly inhaling vapors until it cooled to a reasonable drinking temperature. Terry and Chang had noodle soup for breakfast. I ate two deep-fried egg rolls stuffed with crushed pine nuts and sprinkled with sugar. They were surprisingly good.

Yellow Mountain

After a productive week, Chang and Terry headed back to Shanghai in the Mercedes. I was scheduled to be in Hong Kong the following week, but had the weekend open. So I did something I had wanted to do on several other China trips: Visit Huangshan—China's famous Yellow Mountain.

Huangshan literally means Yellow Mountain. Actually, it's a mountain range located in Eastern China, four hundred thirty kilometers from Hangchou—a five-hour bus ride. Terry had organized the trip for me; it was important to have a good map with all directions and names written in both Chinese characters and English. Fortunately, a private tour reserved by four Chinese couples had an open seat on their minibus. Since one of the men knew Terry, they happily welcomed me to join them. One of the couples, Peter and Judy Wong from Honolulu, spoke English well.

Our group of nine left Hangchou Saturday morning under the care of Mr. Liu, driver and guide. I was ready for some exercise and adventure in the mountains.

Tea plantations were common in the foothills, illuminated by morning sunshine. The new VW minibus provided a comfortable ride with Mr. Liu competently in charge.

"The highest tea bushes on top of the hills produce the best tea. When their young leaves open in the morning, they are harvested; these leaves produce the very best tea," Mr. Liu explained in English and Mandarin as we headed southeast in the general direction of higher mountains and the East China Sea.

Most everybody fell asleep except Peter Wong and me. We struck up a conversation about world travel, with Mr. Liu listening. Peter was in the travel business in Hawaii and was interested in hearing about any extemporaneous adventures I'd had on trips. The one where my exit from a cave in Australia's Northern Territory was blocked by two entangled taipan snakes fighting over a rat was his favorite.

After two and a half hours had passed, everyone was awake. We were at the half-way mark. Mr. Liu asked if we were hungry; a seafood restaurant was just ahead. He spoke in Mandarin but I knew what he'd said; food was pronounced *canyin.* All agreed we should stop.

It was a small local fish restaurant in an unpainted concrete block building, alone without any neighbors. Birds flew around multicolored scallop shells piled high alongside the building. We went inside and sat down.

A big man wearing a rubber apron shouted, "*Huanying!*" (welcome), as he came into the small dining room holding a half-skinned black eel. He waved his other hand for us to follow him.

Another man and a woman were skinning eels out back. Plastic buckets on the ground were full of squirming eels each about two feet long. After making a circumcision cut around their necks, the black skin was pulled off with a pliers, revealing a slender cylinder of pearl-white meat.

A second man was opening scallops with a filet knife. Many of the shells were bright yellow, like the Shell Gasoline logo. He flipped the silver-dollar sized mollusks into a large bowl with the tip of his knife. It was strange; this place wasn't near any saltwater that I knew of—just barren mountain tundra. But then I noticed an old, Russian pickup on the other side of the building rigged with an aerated tank. There must be a connection to the sea somewhere.

Behind the eel skinners and scallop man, another woman was stir-frying eels and scallops together with some root vegetable—probably kohlrabi. Two large stir-fry bowls sizzled over a glowing wood fire.

It was obvious what today's lunch special would be. We were the only customers and what we were observing had to be the day's complete menu. I didn't notice any refrigerators or freezers anywhere.

Samjo, the big guy in the apron, was happy to accommodate us, this seemed agreeable to the couples. Being the invited guest, I didn't argue. We all went back into the dining room, sat down at a long table, and unwrapped packets of wooden chopsticks.

Samjo first brought out individual bowls of white rice—the nucleus of any Chinese meal. Peter asked if anyone wanted a beer; everyone did.

Samjo smiled and went over to a wooden case on the floor containing unlabeled, 600 ml glass bottles of beer—warm, in keeping with Chinese tradition. After he disappeared out back, we toasted each other, clanking the dark green bottles. Like in Turkey, we clanked the bottoms of the bottles.

In short order, Samjo brought out a large tray containing a mountain of food: eel, scallops, and chunks of kohlrabi. The woman who had helped with the skinning brought out individual servings of three different sauces: green, yellow, and brown—the brown being very spicy (*hen la*), mustard and soy. We washed everything down with the warm beer. I preferred the mustard sauce.

Everybody dug in except Mr. Liu; he settled for two servings of white rice—he didn't eat seafood. I noted that the scallops had a slightly off taste and the eel was excessively oily.

Back on the road it didn't take long for things to start happening. Everyone, except Mr. Liu and me, got sick. There was no place to stop with restrooms, only barren country and open road. No buildings, bushes, or trees in the area. Mr. Liu pulled over, and with my help quickly constructed a privacy wall for the women, using their luggage. The men had to organize themselves on other side of the minibus.

Vomiting came first, then diarrhea. With all the doors on the minibus open, luggage wall in place, the ladies relieved themselves on the country side of the vehicle. The men faced the open road on the opposite side without any privacy wall.

It was horrifying for those poor people. I didn't feel at the top of my game, but I wasn't sick. Mr. Liu and I rushed to locate all the water bottles and tissue we could find, handing most of it to the ladies over the luggage wall. We needed a lot more water and tissue.

I won't go into further detail, but you can imagine . . . it was unspeakable. We had to stop three more times and repeat the above scenario before reaching our hotel in Tangkou near Yellow Mountain.

"Food poisoning!" I told Peter who had partially revived. Mr. Liu nodded. The three of us were having tea that evening at the hotel. Everyone else, including Peter's wife Judy, had remained in their rooms since we arrived.

"It was the scallops," I blurted. "Probably dead in their shells for some time without refrigeration." Scallops are bivalve mollusks, unlike gastropods (snail-type mollusks) all of their body parts are contained within their two shells. They were not iced, just submerged in a bucket of water. That should have rung my alarm bell, but I was preoccupied watching the eel skinning—a certifiable live event.

"The eels were all alive; you couldn't miss that!" Peter added.

"Not possible to get sick eating kohlrabi," Mr. Liu commented.

"Indeed, it had to be the scallops," I repeated.

After checking it into the hotel in Tangkou, I arranged for the hotel to deliver eight, five-hundred milliliter bottles of purified water to each of the couples' rooms. A note in Chinese, written for me by Mr. Liu, instructed each person to drink four bottles, (two liters) before morning to flush out the poisons and rehydrate their bodies.

"How come you didn't get sick? You ate the same as us." Peter asked me. He knew Mr. Liu only ate the rice.

I pulled a small amber glass bottle out of my knapsack and handed it to him. The bottle was empty.

"Lactobacillus Test Blend"—Peter read the words out loud. "What is it?"

"A new high potency probiotic that we've developed."

"In Minnesota?"

"Yes, at our company laboratory. It contains billions of Lactobacillus bacteria that colonize the human intestinal tract—and keep it healthy." I could see that put Peter in deep thought.

"It's empty," he said after unscrewing the cap, then handing the bottle to Mr. Liu, who read the complete label.

"I took the last two capsules yesterday morning. It's not on the market yet, so it doesn't have a trade name."

"That is why Mr. Stan did not get sick—the probiotics saved him," Mr. Liu commented, looking at Peter. It was obvious now that Mr. Liu was a man of few words who spoke remarkably clear English. He didn't waste words explaining things that were obvious, but gave clear explanations of things that were not.

"So, large numbers of probiotic bacteria inhibit food-poisoning bacteria?" Peter deduced. I could see he was still somewhat dehydrated but mentally sharp.

"Correct," I replied. Knowing that airport and customs authorities were always wary of such products (white powders in capsules), I had only brought along a single bottle of the Lactobacillus Test Blend. Although probiotics work best as a food poisoning preventative, they also help speed recovery when taken after the fact. I felt bad that I hadn't brought enough for everyone.

The three of us drank more tea and kept talking. The night air was cool.

Peter had mostly recovered by morning and was excited to take the Yungu Cableway with me to the top of Yellow Mountain and do some serious hiking. The others had decided to stay around the base of the mountain in Tangkou for the day with Mr. Liu. Peter's wife Judy insisted that Peter go with me. She felt better but wasn't ready to hike at six thousand feet.

Tangkou is only a few kilometers from the cableway. Mr. Liu gave Peter and me a short lift to the high lift. He apologized for not going with us but was obligated to stay with the others. A Chinese doctor was coming from Tunxi to check them over in the afternoon.

While in the queue for the cable-car, a strange thing happened: A short, gray-haired old man started punching me in the back. At first,

I thought he was just trying to get my attention, but when I turned around, he hit me in the jaw! He had several missing teeth and was grinding the ones he did have while he kept punching. Automatically, I put my right hand on top of his head and held him away. He then started swinging wider, trying to hit me in the chest. Easily a foot taller than him, it felt like I was in a Popeye cartoon. I sucked in my stomach and angled my body away, keeping my hand on his head as he kept swinging and mostly missing.

Peter had been in front of me talking with some students in the queue when he became aware of what was going on. He and two of the students grabbed the old man and pulled him away. A local woman selling cable-car tickets explained something to Peter. She then shook her finger at the old man and repeated, *Zou kai, Zou kai* (go away, go away).

Except for the hit to my jaw, his pounding hadn't hurt. It was the total surprise of not knowing what angered him that got to me. As we waited for the cable-car, Peter informed me that the ticket-woman told him the old man commonly fought with foreigners. He was born nearby, and had never been to the top of Yellow Mountain, not able to afford it. So he hated rich foreigners who could. *Hmm,* I could hear my soul-voice advising me . . ..

And that was the long and short of my first, and hopefully last, fist fight in China.

Peter and I got back in line for the cable-car. It was a Sunday and the queue, surprisingly, was short; we each had packed a backpack for a day of hiking. As we lifted up quickly, I could see the old man trying to pick a fight with another foreign tourist—one shorter than me.

"He's not a happy man," Peter declared. "Socialism does not provide for the pursuit of happiness, only for a state of sameness and conformity." Peter had studied the U.S. Constitution when applying for citizenship. He and Judy were now U.S. citizens; the event with the old man had switched Peter on . . ..

"Socialism does not allow enough winners to succeed and form a successful middle class," he continued. "In socialism a collective hive is formed that operates at a low ebb until it eventually runs out of money and totally fails the people. Then, amidst the ensuing chaos, a dictator elects himself Supreme Ruler and any remaining scraps of democracy are outlawed. Like under Mao.

"My mother was pregnant with me when Mao took control in 1949. My family escaped to Hong Kong just in time. Mother's gold necklace and father's skills as a shipboard mechanic paid our way to Honolulu on an old freighter."

How many times will humanity need to learn this lesson? I wondered. *Socialism has no sustained historical successes. It takes the air out of everyone's tires.*

We watched for the rocks called "Two Cats Catching a Mouse" from the cable-car, signaling our upcoming arrival at the White Goose Ridge Station.

The Yellow Mountain(s) commonly floated in foggy clouds that shifted often to reveal dramatic rocks, peaks and forests. Go into any art shop that sells Chinese paintings and you'll see Impressionistic renditions of the Huangshan area, arguably China's top tourist attraction.

Having only one day, Mr. Liu had recommended we start on the Eastern Steps and be prepared for a good hike—a "seven" on a scale of one to ten. Steep stone stairways defined many sections of the trails. Well broken-in hiking shoes were a must.

The first trail was easy, taking us past "The Flower Growing out of a Whiting Bush Rock" with Stalagmite Peak on our right in the distance. The entire Yellow Mountain range topped off around six thousand feet and it was common to see Huangshan Pines growing out perpendicular from sheer rock faces.

In less than an hour we arrived at the famous Beihai Hotel, a remodeled version of Mao's old communist retreat where he lectured his favorite socialist bureaucrats. I could just imagine them planning what to keep for themselves and what to share with the common folk.

This hotel was strategically located in a high valley surrounded by peaks named Red Cloud, Beginning to Believe, Pen Rack, and Lion Peak—still largely concealed in the clouds at 9:00 a.m.

We stopped for coffee outside a gift shop and checked the trail map. When the clouds started to clear, the peaks magically appeared. Some had pagodas on top.

Several macaques were eating popcorn from a bag they had acquired, probably by theft. It always amazed me how biodiverse the family of macaque monkeys is—these guys were the size of large house cats, brown furred, and caution-curious. Of the twenty-three species of macaques, twenty-two are indigenous to Asia. I couldn't recall an Asian trip that hadn't included macaques—somehow, somewhere. I reminded Peter they could bite.

The trails, many with hundreds of rock steps, were well marked in Chinese and English and we had until 6:00 p.m. to see as much of Yellow Mountain as we could; we didn't want to miss the last cable car down.

The best view was at a place called "Top of Peaks" where it was cold for June due to significant wind chill. I rated the trail to the top a strenuous eight! Both Peter and I could feel it in our legs. But it was worth it—the view was stunning.

After lunch at Mao's remodeled hotel, we hiked to other high spots that were off the main trail, rested at the "Cloud Dispersing Pavilion" for a short time, then off to see Flying Rock—a huge whale-sized, megaton rock standing at an angle next to a granite cliff. A young Chinese boy was trying to push it over the cliff while a woman took his photograph. *Where did the forces come from that put that rock there?* I asked myself.

Anhui Province, Yellow Mountain's home, has a population of sixty million people. To put that in perspective, it's about two-thirds the size of Minnesota which has a population of five and a half million people. I was just starting to appreciate the isolation and expanse of the Yellow Mountain area, the peaceful solitude and outstanding beauty of it, when, at one of the high viewing spots, an Austrian tourist lady handed me her binoculars, and pointed down to a valley between two peaks.

"My God, I can't believe this!" I declared after focusing for my eyes.

"There are thousands of them in that small valley," she said.

"They looked like ants from this distance—ant people in long lines—coming from one place and going to another," I said.

"Look more to the right," She advised.

I turned and could now see the factory. "I count four smokestacks on it." They were swirling gray smoke into the low clouds."

"Workers are changing shifts, thousands of them," she remarked. I handed the binoculars to Peter.

"So much for the pristine solitude," Peter bemoaned.

"The perceived solitude is due to our altitude," The lady asserted. "This is China."

"I bet every valley has at least one mega-factory," I lamented. Peter had a sad look on his face.

When Peter and I exited the cable-car, Mr. Liu was waiting for us with the minibus. I immediately recognized the old man who had been beating on me. He was yelling and shaking his fist at some European tourists. I could see the ticket-selling woman closing up her booth; I ran over to her. At first, she was unwilling to sell me a ticket for the cable-car.

"*Guanbi* (closed)," she mumbled, not looking at me.

"Qing (please)." I handed her 300 yuan ($40): three of Mao's mugshots.

She reluctantly sold me a ticket. Mr. Liu and Peter were standing by the minibus staring at me, wondering what I was doing. I gave them the hang-loose sign in Hawaiian as I went over to the old man who was still yelling at the Europeans.

When he saw me, he stopped yelling. I gave him the two-finger peace sign while I slowly approached. He started walking backwards as I got closer. I held up the ticket and pointed to him. He stopped and stared at me, then at the ticket. In his face, disbelief united with delight while a single tear ran down his right cheek. He stood in front of me motionless as I handed him the ticket and gave him the two-finger peace sign again—my soul voice rejoiced.

Postscript

China's a big place with a lot of people. At over four times the population of the USA on roughly the same amount of land, it appears to be somewhat necessary for it to operate under different mechanisms from those outlined in our constitution. It has been my observation that certain Chinese provinces are ahead of the curve at redefining those mechanisms: balancing strong left-turn tendencies with right turns, rigid controls with more flexible versions. It will be interesting to watch how the young ruling class in Beijing evolves as they replace the old-guard communists. Shanghai, Hangchou, and Yellow Mountain will be good indicators of the pH of the reaction—hopefully more neutral than caustic or acidic.

Author with 155-pound striped marlin caught on
Marlin Boulevard in Ecuador — May 2002

Chapter Seven

Marlin Boulevard
Tales of the Silver Lady

Prologue

When a man is tense and worn from the toils of daily life, his favorite songs don't seem to rhyme, his computer job defines super-boring, and he knows it's senseless to attempt taking the cat for a walk, there is, according to Jimmy Buffet, a solution: Such a man needs to go fishing. Marlin fishing, trout fishing, musky fishing, tarpon fishing, bass fishing, walleye fishing, any-fish fishing—in freshwater or saltwater, an ocean or a lake, a river or a frog pond, it doesn't matter. He simply needs to go fishing. Far and away, with only rare exceptions, it's the most tension relieving, anxiety dismissing, physically conditioning, and authentically exciting, of all human activities. Just don't forget to set the hook!

Salinas, Ecuador May 2002

I was in Ecuador with my sales agent Bernardo Gonzales. We were testing Bacillus probiotics in shrimp ponds and on large banana plantations from Salinas down to the Peruvian border. This involved teaching farmers and growers how to brew (culture) probiotics in aerated plastic tanks so the cost of using high-strength doses (trillions of Bacillus spores) was affordable. Such quantities were required to effectively reduce viruses and harmful bacteria in shrimp ponds and pathogenic fungi on banana plants.

It was a hot Friday in May, one hundred fifty miles south of the equator. We drove along the western foothills of the Andes which run down the center of Ecuador, a band of mountains seventy-five miles wide that bisects the country—its Andean spine.

Bernardo, fifteen years my senior, was a mountain man born in Quito, Ecuador at an altitude of nine thousand feet. At six-foot-four and solidly constructed, head shaved and polished, he looked like Telly Savalas in the 1980s. His *Leggero* tenor voice didn't seem to fit his powerful stature at first, but in combination with his invitingly polite demeanor, these dramatic incongruences helped him disarm skeptical customers. He was a great probiotic salesman.

Bernardo had agreed to let me take him marlin fishing for the weekend—something he had never done. After a successful work week, I was excited to treat him to something different. Being from Quito, where the average daytime temperature is 65° F, he was concerned about 90° F in Salinas. I convinced him that the Humboldt Current running along the coast provided a cool stream of water from the southern hemisphere, which would keep him from overheating while simultaneously activating big marlin. It was a bit of a tough sell, since we had just spent a hot week at sea level in humid farmland. I had to assure him that Juan Perez always kept plenty of cold Mexican beer aboard the *Silver Lady.*

"What's the story behind the name of his boat?" Bernardo asked as we drove to Salinas.

"The *Silver Lady* was Juan's maternal grandmother and, as he once explained to me, she was a kind, elegant lady with beautiful silver hair who raised him from diapers to adulthood. He loved her dearly—she took over for the mother he could never remember. Twelve years ago, she loaned him money to buy the Bertram—his dream boat. She knew he wanted to become a charter-boat captain and chase big fish. She's gone now, but her spirit travels with him and his crew."

"I look forward to meeting Juan and riding the *Silver Lady* down Marlin Boulevard, but I think I'll let you do most of the fishing. I have no experience . . ..”

“Nonsense, you’re going to reel in the first fish, and the third one, and then every other fish after that; we’ll alternate. You’ll appreciate Juan Perez. He’s in his mid-thirties, smart, speaks four languages, and fishing is his life.”

“How many other fishermen will be with us?”

“It’s just you and me, Bernardo! We’ll be the only ones fishing on the *Silver Lady*. Juan and his crew of four will assist us. This is a good time of year for striped marlin; the Boulevard was thick with them last week. When I phoned Juan last night, he told me they had raised fifty-eight over the last three days, had thirty-one bites; eighteen were ‘caught’ and brought to the boat, fifteen were released and three kept. They ranged in size from 150 to 330 pounds—nice striped marlin. The world record is four hundred ninety-four pounds—caught off New Zealand’s North Island in January 1986. They don’t get as big as blue or black marlin but, pound for pound, they fight even harder.”

“When you say raised, you mean you see the fish following the lure while trolling?”

“Exactly. You can see the marlin’s head and dorsal fin following on the surface behind the lure. If the marlin is fooled and ready to eat, it will often slash its bill at the lure, trying to kill it. A “bite” is when they strike or attempt to eat the lure but don’t get caught. A marlin is not caught until you’ve reeled it to the side of the boat and the crew unhooks and releases it—or clubs it and pulls it on board.”

“How many do you think we’ll raise?” Bernardo was starting to show some interest, maybe a little excitement. I’d been with him all week, and finally noticed his extra-large hands—he’d easily handle the big rods and reels.

“Very likely we’ll see quite a few; this time of year Marlin Boulevard is world renowned for producing striped marlin. How many bites depends on how hungry they are and how cleverly we fool them into thinking our lures are something they normally eat. How many are caught depends on how well the hooks are set and how well we maintain control of the fish and keep pressure on it. Also, it’s important to remember to bow the rod to any marlin when it jumps, particularly when it’s close to the boat—this reduces line tension and minimizes breakage.”

“I do believe I might enjoy this,” Bernardo began rotating his right hand as if he were winding a fishing reel. I smiled and told him it was okay to wind the reel on small fish, but he needed to crank it on big ones. When he asked how he'd know which to do if he couldn't see the fish, I told him he'd know. . ..

Salinas is located on Ecuador's Pacific coast about halfway between Peru and Colombia. It's the westernmost city in Ecuador, jutting out on the Santa Elena Peninsula. It has great beaches and is only a hop and a skip to Marlin Boulevard—literally a river in the Pacific Ocean, created by the Humboldt Current that runs from Cabo Blanco, Peru, up the Ecuadorian Coast to Salinas, ending just north of Manta, Ecuador. The Boulevard's cool, low-salinity water creates upwellings of nutrients that feed a variety of fish. Apex predators like marlin take advantage of the rich food chain—it's one of the world's great marlin haunts.

"Are blue marlin found in the Marlin Boulevard?" Bernardo asked.

"Yes they are, but less frequently than striped marlin."

"And they get bigger?" Bernardo spread his arms out.

"Much bigger; the largest caught by rod and reel was a tad over eighteen hundred pounds. Juan has quite a story about a blue marlin—a big one. I'll ask him to tell it."

Bernardo and I checked into the Hotel Blue Marlin del Pacifico near the beach, a sardine's throw from the fishing fleet. I gave Juan a call and informed him we had arrived.

"Have you had dinner yet?" were Juan's first words.

"Nope. We're waiting for you to invite us!"

"Stay where you are, I'll be right over."

"Will do. See you soon."

While Bernardo studied the marlin photos on the hotel's bulletin board, I conversed with the young woman at the check-in desk; she had a blue marlin sewn on her shirt and remembered me from last year.

"Fisherman Stan," Juan shouted when he entered the Lobby.

"Captain Juan!" I answered, smiling wide. We embraced and patted each other on the shoulder—it's called an *abrazo*. Bernardo broke away from the bulletin board, came over, and politely introduced himself.

"You're the mountain man from Quito," Juan declared. "Are you ready for some marlin fishing tomorrow?"

Bernardo grinned. "Now that I've seen proof of your skills in the photos, I'm ready." He embraced Juan with a mountain man's *abrazo*.

While we walked to Juan's truck, it started to rain. As we talked, I could tell Juan was getting Bernardo quite excited—and we were all getting wet. We switched to running.

"Will rain affect the fishing?" Bernardo asked. "It was sunny all day on the banana plantations."

"It's been raining for over a week along the coast; a gloomy drizzle that's driving the striped marlin crazy. We're raising fish—lots of nice fish." Juan and Bernardo spoke to each other in English, not uncommon in Ecuador when an American is along.

"Are they eating the lures?" I asked, to verify what he told me on the phone.

"About half the follows eat the lures—and half of those get off," Juan replied.

"That's par for the ocean." I shrugged.

"At this time tomorrow, Bernardo will be a marlin man!" Juan grinned as he drove through the drizzle; Bernardo was impressed by Juan's confidence.

"Are we headed to the restaurant that serves your grandmother's special dish?" I asked Juan while drying off with a smelly beach towel he kept in the truck.

"Yes, Mr. Stan, we are. By the way, that towel needs to go to the laundry."

"You told me that last September!" I quipped. Bernardo shook his head when I offered it to him.

"Does *encocado de pescado* sound good to you, Bernardo?" Juan asked.

"That's fine. I have it often in Quito, made with mountain trout."

"Here it's usually made with yellowfin tuna but that can vary with the day's catch. Dorado (mahi mahi) and smaller striped marlin are also excellent," Juan explained. The dish is an Ecuadorian specialty: fresh fish cooked in spices and coconut milk is served with rice and crispy *patacones* (deep-fried plantain chips).

"My grandmother's secret was to fry the fish *fast* in coconut oil after they'd been marinated for several hours in her spice mix. We're headed to the local restaurant she once owned."

"*Tres Bohemia cervezas.*" Juan ordered us a good Mexican beer to start things off. The restaurant was full of locals and they all knew Juan. The menu had a photograph of the Silver Lady on its cover—Juan's grandmother, not his boat. She was wearing a chef's apron and holding a platter of her fried fish. Juan told us that whenever he looked at it, he was saddened at first, but then quickly switched to feeling joyful, knowing her spirit was always with him.

While we waited for the three orders of *encocado de pescado* to arrive, I asked Juan to tell Bernardo the story of "The Four Doctors' Fish." I had specifically kept from telling it, didn't want to screw up any of the details.

Juan, after shifting position in his chair, took a long drink of Bohemia. "It was two years ago, almost to the day." He glanced at his watch to check the date. "It had rained the night before, and the morning was gray and gloomy, like it will be tomorrow. The four doctors started drinking beer as soon as we left the harbor at 7:00 a.m. We headed out directly west along with a dozen other charters.

"I don't concern myself with the ocean's depth anymore; it varies by thousands of feet from place to place and doesn't tell us where the marlin are. They follow the Humboldt Currents' upwellings, which are higher than sea level and bring up plankton that feeds concentrations of baitfish—sardines, anchovies, jack mackerel. These become feed for Bonita and yellowfin tuna which then become feed for the marlin. It's a tight, unique ecosystem."

"How do you rig your fishing gear to catch such large fish? A two-kilo mountain trout was the biggest fish I ever caught." Bernardo remarked. I knew this would enliven Juan exponentially. He loved to talk fishing gear and techniques as much if not more than I do.

"We were trolling fourteen-inch, skirted baits with black and blue plastic tentacles, my favorite gloomy day color. They looked like squid. We had six baits out, each rigged with a 10/0 hook on a fifteen-foot leader of four-hundred-pound test monofilament; the leader was attached to nine hundred fifty yards of one hundred thirty-pound test monofilament on Penn International reels. I varied the trolling speed from eight to fourteen miles per hour, trying to stay on the edges of higher water upwellings, looking for schools of baitfish on the sonar. Just like we'll do tomorrow!

"Several hours passed without any action. The doctors were well into their beer and playing cards up on the bridge. The day brightened somewhat at noon and we ate lunch—subs, chips, and beer; Inca Cola for me and the crew. I was half done with my sandwich when the marlin ate the outermost bait on the starboard outrigger. Never saw it coming. . . In .seconds all hell broke loose.

"It was the biggest blue marlin I'd seen in ten years of chartering these waters. On its first jump, it completely cleared water by twenty feet before falling back into the ocean making a splash higher than the tuna tower—a splash that, in my experience, only a breaching humpback whale could make. The marlin jumped and raced four hundred yards of line off the reel before Joseph—the bone doctor—could maneuver into the fighting chair. Luis, my lead mate, set the hook three times, then adjusted the reel's drag and held the rod until Joe was ready. Instead of playing cards, Dr. Joe should have been sitting in the fighting chair all along—the rule for the man up next to fish.

"Line screamed off the reel! After the doctor grabbed the rod, he put the end of it into the chair's rod cup and started reeling and guiding line on the reel with his left thumb—he seemed to know what he was doing. The crew brought in the other lines; it didn't take the doctor long before he was straining like a pack mule struggling uphill in the Andes." Juan smiled.

"He had to crank the reel, not wind it." I winked at Bernardo

"This fish had unbelievable power. After an hour of acrobatic jumps and grey hounding over the surface, it went deep and pulled hard. For the first two hours all we could do was stay even with it," Juan shook his head. "The four doctors took turns on the fish, each one lasted for about an hour."

"So, this is where the story really starts, I interrupted. Pay close attention Bernardo, we don't want what you're about to hear, happen to us tomorrow," I emphasized, just as the food arrived.

Fresh yellowfin tuna was served on a bed of long grain white rice. The aroma was a tantalizing blend of coconut with traces of basil, rosemary and red pepper; it didn't disappoint. It was superb.

As we ate, Juan continued to explain in detail what happened during the four hours it took to get the monster marlin to the boat.

"Was it muscle strain or courteous professional sharing that got them all involved?" Bernardo asked Juan.

"It was the beer! Too much of it—not one of them could have pumped and cranked for four hours. I had to follow that fish with the boat, faster than I normally do, or we'd still be there pumping and cranking.

"Like I said, during the first hour of the fight, the fish jumped frequently—tremendous jumps that completely cleared the water before smashing back into to sea. That's when I knew its great size, all of sixteen feet long and thick with girth through to its tail. When it dove deep, it could run out a hundred yards of line in four seconds—about eighty miles an hour. While following the fish with the boat, I had to adjust speeds to make sure a doctor could reel fast enough to keep up and not allow the line to go slack. I told Luis to reset the lever drag on the reel tighter, or we'd have run out of line. All my rods and reels had been re-spooled with fresh line."

Juan ordered three more Bohemia for us.

Talking about this fish seemed to put Juan in a trance. It looked as though he was back in time driving the Bertram, reliving the story, recalling his precise orders to the crew and their responses—but also adding a narrative of action and details necessary for us to understand the event. Later he told us it was like watching a DVD recording of the past, only in three dimensions— like a hologram. Bernardo and I were spellbound.

Juan's continued recount of catching the huge blue marlin:

> "Pull together! Synchronize!" I shouted to my four crewmen as they pulled on the head of the dead fish,

which was hanging over the port corner of the transom. I backed the boat slowly into the choppy sea to reverse the fish's inertia which was pulling it back into the water. If the crew let go, she would sink two thousand feet to the bottom.

"She's a slippery giant," the elder Kichwa native shouted in Inca talk. Slippery from slime that had thickened during the fight, the dead fish sloshed rhythmically with the waves.

"Put gloves on, dammit," I yelled back, steering from the bridge above them. Most of the colossal marlin was underwater, only its gargantuan head with its bulbous eyes and immense bill—a lethal sword—were on the transom. The rough sea wanted its prized marlin back. We had to boat this fish! I kept steering at an angle to the incoming waves to lessen the slippage. I shouted orders only when necessary. My men were experienced.

The two Kichwa men each grabbed a pectoral fin and half of the dorsal fin on opposite sides—straining to pull the fish forward into the boat. The two younger mestizo men gripped the bill and pulled. The magnificent blue iridescence of the fish was gone, as was the shine of her *amarillo* lateral line— replaced by a blackened gray. Fighting four fishermen in addition to a thirty-three foot boat was too much for her; she expired on her last run just before Luis grabbed the bill. She instantly became dead weight. The crew got the head pulled onto the transom—that's the best they could do before the sea took over and pulled back. The transom had no gate to open for landing such a fish, like those on some newer model fishing boats. I wished that my boat had such a gate. My mind was going crazy—I had to get back to rational thinking.

The Bertram's transom was five feet above the water line and the fish had to be pulled over it. The men were yelling orders to each other in Kichwa and Spanish: "We can't lose this fish!" they kept repeating.

"Use the long gaff, get more leverage," I yelled down from the bridge, standing with my back to the wheel, steering with one hand.

Luis, the taller of the two mestizos, grabbed the gaff and spiked it into the fish just behind its head—then shouted a cadence that coordinated the pulling. The gaff seemed to help; they were able to pull in twelve more inches of the fish.

The four doctors had been watching the ordeal from the bridge, drinking beer and arguing who had cranked in the most line during the four hour fight. There was room for one of them to help pull in the fish. I pointed to Jay Petersen, another bone doctor, to go down and help. Clumsily he missed a step rushing down the ladder—another miss and he would have required his own services. I had to get down and help too, we had to get a rope on her—worst case, we could drag her back to Salinas.

"Joseph! Take the wheel and keep on a heading of two-hundred-seventy degrees west. Don't touch the throttles," I hollered while sliding down the ladder.

The men gained another foot on the fish with Petersen helping. I went to the transom and leaned over to get a better look—the marlin was way more than a grander, meaning it weighed over one thousand pounds. This blue marlin had to weigh close to two-thousand pounds. I could do the math in my head for calculating weight from its measurements.

I knew the All-Tackle world record was 1,376 pounds, caught by following strict IGFA (International Game Fish Association) rules specifying one angler. The largest ever taken by rod and reel was 1,805 pounds and the largest ever caught long-lining, commercially, and brought into the Tsukiji Fish Market in Tokyo weighed 2,488 pounds.

If my estimate of two thousand pounds could be confirmed on a certified scale in Salinas, it would be the largest blue marlin ever caught by rod and reel. *How famous would that be for the Silver Lady!*

I glanced back to see the five men pulling on the corpse of the great fish. *How many oceans had she explored, how many attacking sharks had she lanced with her bill, how many high-speed escapes from pods of orcas, how many juvenile blue marlins had she spawned?* Those were just some of the questions that crossed my mind.

Had the marlin not died from the extended fight, I would have insisted it be released. I was sure of that now. I recalled my mind laboring when I chased the marlin with the boat, making it possible for the doctors to pump it up and reel it in. *Kill it or release it,* flashed on a billboard in my head. The dead fish would be proof of my expertise as a charter guide and captain. It would hang high on the dock in Salinas and amaze the crowds. But after the four hour fight, when it finally came to the boat, totally exhausted from an unfair fight, it died. I didn't have to make the difficult decision, the fish had. But now the soul of the sea was taking revenge, trying to take its great fish back.

Sorrow griped me for some moments. But then my mind reverted to the realization that it was just a fish, and catching fish was my livelihood, my purpose in life. My thoughts settled down: *It would certainly prove to be largest marlin ever caught by rod and reel and the meat would feed many local people.*

Suddenly the wind's velocity picked up and most of what had been gained on the sixteen foot fish had slipped back into the sea. Luis shouted, 'We're going to lose her!'

"No!" I yelled. "Jorge! Go get that heavy nylon rope from the cabin and bring it to me."

Jorge, Lewis' brother, sensed a challenging assignment. He was the youngest and best swimmer of the crew— longhaired and thin, strong for his age.

I tied one end of the rope to a cleat welded to the boat's mainframe and gave the rest of it back to Jorge. I told him to, 'Jump in the water and tie the other end of the rope to the fish's tail—three loops around the tail then tighten with a five-wrap uni-knot. You'll have to dive down ten feet."

"Yes, captain." Jorge saluted and jumped off the transom with the rope. He followed my instructions and was back in the boat in five minutes.

"Jorge, shaking his long hair, repelling the sea, had a frightened look on his face. "Did you pull the knot tight?" I asked him.

"Yes, captain."

"You look frightened."

"There's a monster shark following us, much bigger than the marlin." Jorge was shaking.

"Was it attacking the marlin—eating it?" I asked him.

"No, but I think it will."

"'Good thing you're back in the boat!" With no time to waste, I shouted up to the other two doctors to get down and help pull in the fish, there was room now with the long rope. Joseph stayed on top, steering.

"Is it a great white?" I asked Jorge.

"No, its teeth are slanted sideways, and it has stripes." Jorge shivered—the wind had picked up.

"Jesus, help us. . .," I mumbled. "Everyone pull together! Synchronize!" I yelled to all the men. The marlin began to inch forward again—just as the sea exploded.

A ton of seawater instantly drenched everyone as a monstrous, seven-finned, bronze torpedo engulfed the underwater portion of the marlin. Leaping out of the sea and scraping the side of the boat, it pulled the entire two thousand pound marlin carcass loose from the grips of seven men. Seconds later, the rope tightened and jolted the boat hard. The massive head of the marlin—dead but still wielding its lethal bill—just missed two of the men when it was pulled off the transom into the sea.

I jumped away and raced up the ladder to the bridge and grabbed the wheel, Doctor Joe was on the floor.

"Stay on the floor!" I told him.

"Everyone hold onto something!" I yelled as I hit the throttles on the twin 425 horsepower diesels, vibrating everything. The port corner of the boat pulled down to sea level from the weight of the marlin and the gargantuan bronze shark eating it. Both were in the water directly behind the boat. More water poured in.

"Cut the rope!" I shouted. Doctor Petersen, totally sober now, did just that, completely releasing the marlin carcass.

All eyes were on a tiger shark of unimaginable dimensions eating our two thousand pound, sixteen foot, blue marlin. The shark's giant head slashed violently back and forth, ripping off layers of marlin meat. Dagger-sized, serrated teeth pointed sideways, backed up by extra teeth folded behind the ones doing the ripping, all crimson red with marlin blood. The animal's recessed black eyes—each larger than a soup bowl—directed the voracious feeding.

The boat leveled out after the rope was cut and water gushed back into the sea thanks to two oversized bilge pumps. I didn't want that tiger shark jumping into my boat—it was very close to doing just that after engulfing the marlin's midsection.

The two native mates were talking feverishly in Kichwa. I listened to them as we all watched the giant alien eat. "It follow a long way—even before marlin die. Orca-size tiger shark very rare. We see one long time ago—it weigh four thousand pounds," they confirmed in Kichwa.

"Holy blue-footed boobies," Doctor Petersen exclaimed. He'd been to the Galapagos Islands and knew booby birds. Nobody laughed. None of the doctors had any conception of what was normal size versus gargantuan when it came to marlin and sharks. There were six thousand pounds of meat in the water directly behind my boat—and two thirds of it was alive!

"That's what I saw when I tied the rope around the marlin's tail," Jorge stammered, pointing to the monster tiger shark that was ripping apart the largest blue marlin ever caught on a rod and reel.

Juan took a long drink of his beer—I followed and so did Bernardo. Then I took a large cotton napkin and wiped the sweat off Juan's face and neck. We just sat in the back of the restaurant and stared out at all the locals eating fish. We sat quietly for some time. . ..

"Are you okay?" I asked Juan.

"Yes, sorry. Wow, I really got into it."

"You sure did," Bernardo added. "The story took control of you!"

"How long did it take to get back to Salinas?" I asked, handing him the napkin.

"It took us two hours to get back running at twenty-two knots, plenty of time to talk. Joseph, Jay, and the two other doctors did more listening than talking," Juan explained. "It was clear *then* to everyone that what they had experienced was extremely rare."

"It was amoral nature in the raw; the product of three-and-a-half billion years of evolution from algae to hominoids, with a few stopping-off points along the way, where, after achieving perfection in functions, certain organisms advanced further only by increasing in size. A two thousand-pound blue marlin was one example, a four thousand-pound tiger shark another. Even Las Vegas computers couldn't put odds on the happenstance of two such rarities coming together," I added.

"How rare is such a tiger shark? Can they get bigger still?" Bernardo asked.

Captain Juan explained what two New Zealand clients had told him: "They ran a deep-sea guide service in Bali. One was an ichthyologist that specialized in sharks. He had witnessed unimaginably large tiger sharks in the Indian Ocean; one was twenty-three feet long and weighed fifty-two hundred pounds. It had been caught by a commercial Japanese trawler and was two hundred pounds heavier than the largest known great white shark.

"The Kiwis had left me with some advice: Tiger sharks never stop growing and have an advantage over great whites due to their feeding habits, which are more diverse, aggressive, and ravenous. Never underestimate how big they can get."

We had great fishing the next day. Bernardo got hooked on fishing marlin and big wahoo. He got a fifty-pound wahoo as the first fish of the day before I got a 155-pound striped marlin on twenty-pound-test spinning tackle. The marlin ate a ten-inch, turquoise-colored, Big-Jake musky bait that looked just like a wounded mahi mahi in the water. Juan covered his eyes. He didn't want to believe a marlin would eat a musky lure. After that, Bernardo caught a 250-pound striped marlin—a beauty—on one of Juan's black and blue skirted squid baits. All told, we had seventeen follows from striped marlins, caught nine and released all but one—the 155 pounder I caught that Juan kept for eating. Both Bernardo and I had sore arms, backs, and wrists. I shared a large tube of Bengay with him that night, sanctifying the entire second floor of the hotel with methyl-salicylate vapor.

"Still want to do this again tomorrow?" I asked him before hitting the hay.

"*Si, Amigo. Si*! For a little guy, you set the hook good in me."

I embraced him and patted him hard on the shoulder—attempting it mountain-man style so he could feel it.

Author by Rainbow Bridge outside Forbidding Canyon —
Southern Utah, 2002

Chapter Eight

Forbidding Canyon
The Secret of Nonnezoshe

Prologue

"The spell of the canyon comes back to me, as it always will come. I see the veils, like purple smoke, in the canyons, and I feel the silence. And it seems that again I must try to pierce both to get at the strange wildlife of the last American wilderness—wild still, almost, as it ever was." Zane Grey, 1915

Kayenta, Arizona—April 2002

I was filling the Yukon with gas and enjoying a conversation that had me as the center of attention. It all started when I drove into the station and asked if anyone knew a good trail guide. A Navajo man with shoulder-length black hair spoke up. "I might be able to help. Where do you want to go?"

"Do you know the old trail to Forbidden Canyon where Zane Grey's party came through in 1914?" I asked him.

"Like the back of my hand!" he asserted. "I even know the cliff where Teddy Roosevelt took a pee in 1913."

"That's not on my bucket list, but. . .."

"I can take you there for three hundred dollars. My name's Ki' Somma, I normally charge tourists four hundred." He was a forty-something, clear-speaking, dark-skinned Navajo. I later learned his name meant *The Sun.*

"On horses for three nights?" I held up three fingers.

"We can do it in two nights—thirty-one miles round trip."

"I'd like an extra day, so there's time to explore; I'm a rockhound."

"Okay, we can camp three nights—need to add fifty dollars." I understood, nodding, just as the gas pump shut off.

"He's not a licensed guide," a younger Navajo man with braided hair commented as he finished washing my windshield.

"No matter. Ki' Somma knows the canyons and rocks better than any guide," an older, Indian man in a grease-spotted tee shirt declared as he sat changing a tire.

"He's the *spirit* and he knows!" Ki' Somma pointed to Achak, the older Indian.

"Ki' Somma's mustangs will be hard to ride; our ranch horses are easy riders." Chetan, the younger Indian, pushed the braids out of his eyes. "My name means *hawk,* and I'm a licensed guide!"

"Can you ride a horse?" Looking at me and ignoring Chetan, Ki' Somma wiped grease off his hands,

"I can ride a horse."

"Then you won't have trouble riding any of mine—Iberian mares bred with wild- mustang stallions . . . three generations ago. I personally geld colts at six months and break them at three years. They know the canyons and how to negotiate slippery rocks. But I've got a very special six-year-old mare in mind for you!"

"I'll only charge two hundred fifty!" Chetan offered, interrupting again.

"Chetan!" Achak elevated his voice, adding something in Navajo—then pointed to a plastic chair next to the service station's door which signaled Chetan to go sit down. "He's Ki' Somma's nephew, my grandson. There's competition in the family—capitalism.

"A better description is, 'economic freedom,'" I said. Achak smiled.

I grinned and shook Ki' Somma's hand. "Let's do it." I walked over to Chetan and gave him three dollars for washing the Yukon's windows. "Don't be so quick to change the price. Emphasize your youthful energy instead. You'll have other opportunities; economic freedom creates many."

I'd now met the *spirit, hawk,* and *sun* in Kayenta, Arizona—with plenty of day left to hit the trail with *the sun.*

I followed Ki' Somma's truck on US-160W eighty miles to Navajo Mountain, Utah. At over ten thousand feet, Navajo Mountain is the highest peak in the Navajo Nation. Below its snow-capped top, spring-green pines, junipers, and incarnadine-rock extrusions clearly define it—it's easily recognized from a distance. The name of the mountain in Navajo, *Naatsis'aan*, means "Head of the Earth Woman" and is vital to the Navajo's creation story.

As we entered the town of ninety-five households and three hundred and fifty residents, Ki' Somma found his sister grilling venison hot dogs along the highway with her big Navajo family. She invited us to join them for lunch. He and I, both starved, sat on the ground and ate hot dogs and drank warm Pepsi. He outlined the "specifics" of our trail ride: We would leave after lunch—this lunch! Then camp out for three nights on a mission to discover the secret of the Forbidding Canyon and Nonnezoshe. The details would invent themselves!

Two and a half hours earlier, I'd pulled into that gas station in Kayenta, hopeful I would find something to do for the rest of the day—and fortunately I did. This extemporaneous mode of operating has suited me well.

Ki' Somma and I discussed what food we needed to pick up in town and anything else I might need—which was nothing! I had prepacked for several overnighters in western canyons before leaving Minnesota. It had taken a couple of decades to learn that less is more when hitting the dusty trail in the American West.

After buying the groceries, I followed Ki' Somma to his sister's octagonal hogan, where I parked the Yukon while he readied the horses. He told me to leave my cooking gear; he had what we needed. I put on a multi-pocketed vest while he changed shirts. The tempo was fast, no monkeying around. He had the horses ready in a short time.

"Try out Madonna, she's the very special mare I mentioned—six years old and loves canyon wildflowers—knows the ones safe to eat. See if she takes to you." Madonna was a stunning white Iberian mare. Ki' Somma handed me the reins. It was a test. I knew what to do.

Standing next to her, I held the reins and gently patted her neck while introducing my energy aura, then moved to stand in front of her. "Hello, Madonna, my name is Stan. You are one beautiful horse—smart too, I hear," I said in a soft voice. "I'm excited about our ride together." She nickered nicely. Ki' Somma grinned as I mounted her. My two hundred pounds got her attention. I remained still for a couple of minutes to allow her to acclimate to me. Then she slowly walked over to a water bucket, glanced back to make sure I was upright, and took a drink.

We headed out doing an easy warmup trot, riding through the desert northeast of Navajo Mountain. I had a lightweight backpack, sleeping bag, and mat tied to the saddle, plus half our food in my saddle bags. I followed Ki' Somma, who was riding a five-year old gelding named Tonto, with the rest of the food, cooking gear, and a two-man tent. Ki' (pronounced "kee"), as I called him with his approval, told me he would normally sleep outside the tent, unless it rained. For a three-nighter we were lightly equipped. I pulled down the brim on my leather Stetson and spoke encouragement to Madonna. I was already in sync with her rhythm.

"Does the Navajo Nation bother you about not having a guide's license?" I asked Ki.

"No. As long as there's no paperwork, no problem. I'm not a guide, just a Navajo man. You bought your Navajo Nation camping permit in town. That's what's important. We're friends, and you've invited me to go camping. The three hundred fifty dollars you paid was a gift for my birthday—which is next week."

"Happy birthday, Ki'." I saluted him. "I approve of your business adjustments, at least out here in the Wild West."

"The Navajo Nation is happy they don't have to support me." He grinned. "If I believed in socialism, I'd be lying back in a hammock swatting flies with all the time in the world on my hands . . . waiting for somebody to buy me a beer."

I laughed. "Don't start me on that!"

"I hear you!" He picked up the pace now that the horses had warmed up.

Nine miles from Ki's sister's hogan we came to the rim of Cha Canyon—elevation 6,400 feet. In the distance to the northwest, across the waters of Lake Powell that we couldn't see, was the beginning of the Grand Staircase Escalante National Monument. To the northeast, giant red mounds of sandstone—jasper-colored in the shade, bright vermilion in the sun—stretched for miles until they connected with the San Juan River. Big country.

This was the Wild West circa 2002, exactly as it was in the 1860s when Kit Carson and company failed to capture the Navajo's leader Hashkeniinii and his followers. We rode down into Cha Canyon on the sage-bordered trail; brown lizards scattered with our approach. These canyons out here often hold the remains of hogans and sweat lodges from the Navajo's past. This desolate area along the Arizona-Utah border hadn't changed in millions of years, and then only where fast-moving, stone-cutting rivers pierced the canyons. Looking ahead, I appreciated the massive upheavals of the distant past and the irreversible erosion that fast water had caused.

This would be our shortest day of the three; it was already 3:00 p.m. We would ride until dark, then camp before the greater maze of canyons was upon us. Ki' had alerted me to expect continual changes in elevation: up five-hundred feet then down five hundred feet, up again then down again . . ..

The horses were very observant of the trail's condition at different elevations, often re-positioning their legs to minimize slippage. They had to step around a giant's grab-bag of rocks, large and small, many loose. I was thankful that I wore nylon support bands on my knees—it helped them endure the bumps and jerks. However, nothing could stop my XL bum from bouncing—a good filming opportunity for a Preparation-H commercial.

"Are the horses shod with steel shoes?" I gently prodded Madonna to keep up with Tonto.

"No! They would slip and slide like crazy. Both horses have been shod with polyurethane shoes and anti-slip rubber studs for maximum grip and minimum slippage on canyon rocks."

I smiled to myself. *Plastic horseshoes—a technological breakthrough.*

"Don't override Madonna when we come onto slickrock (steep, smooth rock surfaces). She's smart, and like Tonto, knows where to go slow and when to slide if necessary. In certain areas we'll walk the horses. It's easy to lose the trail in these canyons. Stone cairns usually provide guidance, unless 'licensed guides' have knocked them over!"

We slowed down to rest the horses and let them nose through a patch of purple sage, like Zane Grey wrote so eloquently about in 1915. We followed much of the same ground Grey covered in his acclaimed novel, *The Rainbow Trail.*

"Madonna keeps right in step with Tonto," I remarked.

"She tolerates him well—knows he's been gelded and is harmless."

"Do you ride her?"

"Yes, around town in Naatsis'aan. Sometimes I alternate with several geldings I'm breaking in. Tonto is only five years old; he's been a fast learner. He's a cool, calm, collected horse with no testosterone episodes." Ki' patted Tonto just as the horse looked ahead and whinnied. Madonna followed suit.

"Yes, Tonto. Yes, Madonna. That's a creek you see," Ki' assured them. We stopped to let them drink beside a large patch of prickly pear cactus in full bloom with hot-pink flowers.

"Watch Madonna eat the flowers." Ki' pointed to the cacti. She carefully pulled them loose with her lips, without getting pricked! Tonto was a bit more vigorous with his effort, and got his tongue stuck with thorns. Madonna short-snorted him—a warning to be more careful. Ki' removed the thorns with a pair of pliers.

As we moved on, we came upon an extensive patch of red Indian paintbrush. Madonna cut in front of Tonto and wouldn't let him go near it. Ki' would have stopped him if she hadn't. "Very poisonous for horses," Ki' asserted.

"Probably due to its high selenium content," I commented. "A necessary trace mineral that's poisonous in very small quantities—only three milligrams (mg) per day is recommended for horses."

After an hour of weaving our way down the uneven trail into Cha Canyon, we then came back up and out of the canyon on multiple switchbacks. I tried to keep my body parallel with the trees as much as possible; they grow straight and don't lean on angles. It's important for riders to keep their weight centered on the horse; leaning forward when going uphill puts too much weight on the horse's front legs, exactly the legs it needs to climb with. Going downhill, leaning back puts too much weight on the hind legs that control descent. For moderate slopes, a rider's leaning should be barely perceptible.

The horses continued to impress me. They carefully studied the trail whenever it became precarious, which was most of the time. When negotiating the challenging switchbacks on smooth slickrock, I could tell the rubber-studded, plastic shoes made a difference for the horses. Going up and out was slow, but sure.

Rugged wilderness surrounded us. The afternoon sun's rays exploded on the giant rocks, producing a brilliant redness. The canyon—from Time's Basement—was windless, silent except for the noise of our clumsy invasion. A good shout produced exceptional echoes.

Once out of Cha Canyon, we followed an escarpment of sandstone cliffs and mounds until we came to another formidable canyon.

"Bald Rock Canyon." Ki' declared, pulling up on Tonto's reins to give the horse a better view of the stratified plum-colored sandstone that appeared to be blocking our way. Numerous circles and curves of it stacked up—deep gouges separated the stacks. Madonna snorted as she looked down and around. Both horses studied the strata. Ki' gave them several minutes to do this before we forged ahead. Then, slowly, keeping parallel to a ten-foot-deep gouge on our right, Ki' guided Tonto left. Madonna followed. It was hard to comprehend how this could be part of the trail.

"There are two types of sandstone here: the hard, compressed Kayenta variety and the softer, less-compressed Navajo form. Kayenta formed the base and walls of the red canyons, the pinker Navajo type composed the trails and rocks closer to surfaces," Ki' explained. Each presented different challenges for the horses. Monster Jurassic rocks and deep alcoves awakened the geologist in me. This place alone would take a week to explore. We moved on in the direction of the oncoming sunset.

Ki' pointed to several rapacious vultures flying over us. "See the dead burro below?" He pointed down. "There's a drove of burros around here—wild ones."

"Looks like this one was gutted by some critter," I remarked.

"Probably coyotes!" Ki' twisted in his saddle looking around. "Those burros are too smart for their own good. When in danger, they take time to assess the situation before running."

"What about wild mustangs?" I loosened my hat; April sunshine had warmed things up.

"Lots of them in Nevada's Great Basin. Here in Utah, too. They run like hell at the first sign of danger. They're not considered to be as intelligent as a burro." Ki' smiled, reaching into a saddle bag to pull out two green apples. "For the horses." He tossed one to me for Madonna. "If you ever want to get a western trail horse to smile, give it an apple after riding it up and down a desolate canyon.

"There are creeks from snow runoff in Nasja Canyon and Surprise Valley, we'll camp by one for the night. I wanted you to get a sample of canyon action today, they'll be plenty more tomorrow." Ki' smiled and picked up the pace.

I took off my Stetson, held it high, and yelled "He-ha!" as Madonna robustly followed Tonto.

Fading sunshine illuminated the slickrock 'benches' between Bald Rock and Nasja Canyons. When we rode into Nasja Canyon both horses perked up at the expansive view. Sheer, red-rock cliffs dominated.

Ki,' of course, knew of a good place to camp that had firewood—often a scarce resource—and some green treats and water for the horses. I pitched the tent while Ki' built a fire. Fresh corn on the cob and T-bone steaks from northern Utah's grass-fed cattle were on the menu—I did the grilling. Our next dinner would be less elegant: freeze-dried spaghetti with freeze-dried peaches for dessert.

After dinner Ki' unsaddled the horses and let them loose to graze, first saying something to Madonna in Navajo. I didn't need an interpreter. His message: "Keep an eye on Tonto, don't let him wander too far." I swear I saw the horse nod.

Ki' positioned the saddles near the fire and stretched out with his head on one. I reached into my backpack and located a stainless steel flask containing Chambord liquor. I opened it and handed it to Ki.' He took a drink, raised his eyebrows, and handed it back to me. Before taking a drink, I raised the flask and looked him in the eye. "To the quintessence of a thousand raspberries preserved with ethyl alcohol. The 'spirit' of the gods."

Orion's nine stars, forming the image of a great hunter aiming his bow into the Milky Way, shone above us.

As we kept the fire going, Ki' was happy to share some of the fine points of Navajo beliefs. He started out with a quick review of the Navajo Creation story, detailing the first three worlds that preceded the Fourth World, our current time: "The First World was black-dark with four seas, and in the middle there was a single island with a single pine tree where all the ants and other flying bugs lived—the Air-Spirit People. The seas were ruled by supernatural beings, like giant frogs. Here is where colored clouds came together to form First Man and First Woman. In the Second World, various birds and other animals came into being, and sexual activity advanced. The Third World is where the holy people lived—they were immortal and traveled by following rainbows. Here, First Woman gave birth to five sets of twins that reproduced and increased the population. Things brightened up in the Fourth World when the sun and moon were created."

"Do you believe that?" I felt I knew him well enough by now to ask.

"There are some lessons of evolution in it . . . bugs came before frogs, and birds came before people. But, no, it's a Navajo myth—good campfire talk."

"What do you *really* believe?" I asked him.

Ki' paused for a few minutes, looking up at Orion. "Balance is key. The Great Father Spirit doesn't control good and evil down here. It goes in cycles for the Navajos as it does for everyone. There are good times and bad times. You have to know when to hide and be still, and when to be loud and play the game." He looked into the fire's embers, then stirred them with a stick. "This doesn't mean good and evil must average fifty-fifty to be in balance—time curves and twists differently for people. Keeping the good experiences above the fifty-percent mark, and holding them captive for as long as possible, is the secret to a happy life." There was a bright light within this dark-skinned Indian man. His voice was intelligent and hypnotic.

"I agree with you," I conceded. "You have captured the essence that other religions seek to find."

"What do you believe?" Ki' gave me a genuine look of curiosity.

"I've travelled the world—all of its continents except Antarctica—and have had many conversations with people of different faiths. Except for Christianity, a common denominator for most religions is that the master creator is ill-tempered and threatening.

"I was born and raised Catholic and couldn't understand why the Vatican's philosophy, also ill-tempered and threatening, so opposed Christ's. With this enigma haunting me, I searched the world for clarity.

"What I have discovered, unquestionably, is that there exists a much broader reality which surrounds and penetrates everything. A reality so immense and complex that we humans don't come close to having the necessary sensory abilities and intellect to fully experience it. It's spiritual and interdimensional, microscopic and macroscopic, energized by light and, strangely, by darkness. In this reality, time is an illusion.

"Staying on the upside of your balance equation can be difficult for many people. Jesus Christ understood this and the special insights and qualities that only mortal humans possess. He volunteered to come to Earth, live as a human being, and agreed to die by crucifixion, nailed to a cross for six hours. Both to qualify for our belief in him and to absolve our sins. I don't have to believe this anymore; I simply know it's true.

"So Ki'—here's the solution to life's equation: I've seen it only once and then only through a tiny keyhole in a giant door. There is a divine trail that runs through this reality. A trail that pierces the dimensions and complexity and delivers deserving humans to a wondrous and simple realm . . . and Jesus is our master guide."

On that note, we both took a long drink of Chambord.

Wearing a sweatshirt, I slept on top of my sleeping bag in the tent, cool but not cold. Ki' slept next to the fire on a wool blanket with his saddle as a pillow. Morning came with an olfactory signal: Bacon, eggs, and coffee. Ki' was up cooking.

Madonna had apparently kept Tonto under control as far as I could tell. They were both here, eating grass behind the tent. After breakfast, we saddled up. Ki' checked the tightness of my saddle's straps, while Madonna wiggled her hips.

"Okay, good," he approved. Happy I'd cinched the straps correctly; I mounted and patted her neck. . ..

"Over here!" Ki' called, he was on Tonto near several close cedar trees. I gave Madonna a slight nudge with the heels of my boots, she responded.

Ki' pointed to four large paw prints in the sand between two of the trees. "Big cat!"

"Cougar," I claimed. Easy to recognize rounded triangular pads each with four, wide, tear-drop claws—large claws compared to ones I've seen elk hunting in Colorado. Smart cats follow hunters hoping to find a gut pile at the end of the day.

"Eight feet long, two hundred fifty pounds," Ki' affirmed, rocking back in his saddle with his arms stretched out to mimic its size. "Last night, the horses were away grazing, or they'd have gone crazy."

"Is it common for cats to get that big in this area? I know the ones on the north rim of the Grand Canyon get big by eating elk. They can attack an elk six times their weight."

"Before the sheep ranchers came, it probably was rare out here. It would have been hard for a cat to get big eating lizards and packrats. The male cougars I see around here are big. We'll keep the horses tied up and closer tonight—they're the best cougar alarm.

"Don't worry," Ki' pulled a Colt-45 with a seven-inch barrel out of his saddle bag and held it high. "I sleep well with this," he announced. I didn't need to show my snub-nose Ruger 357.

We rode through Nasja Canyon toward Surprise Valley that morning to places memorialized by Zane Grey. Here's where we had to walk the horses to get down into the valley, guiding them through huge sections of broken cliff and rubble from a rock avalanche that led to Surprise Valley. The valley was rich in biodiversity, more so than the rocky canyons. More water equals more biodiversity.

Here's a passage from Zane Grey's novel, *The Rainbow Trail,* describing what the protagonist, Shefford, saw when he first entered Surprise Valley: "He faced a wide terrace, green with grass and moss and starry with strange white flowers, and dark-foliaged, spear-pointed spruce trees. Below the terrace sloped a bench [rock formation] covered with thick copse, and this merged into a forest of dwarf oaks and beyond that was a beautiful strip of white aspens, their leaves quivering in the stillness. The air was close, sweet, warm, fragrant and remarkably dry."

We spent some time in the large valley. The April sun didn't disappoint; it was pleasantly warm and dry. While walking Madonna, I found some clear quartz crystals near the ruins of an old Hogan—a small home for a Navaho family "many moons ago." Now it was hardly more than a square wood pile; the clear rhombohedral crystals were probably collected by a family member. I kept one that was six inches long and double-terminated.

"That crystal will concentrate spiritual energy," Ki' revealed.

"Smart man!" I pointed it toward him. "Crystals also promote sharp memory and sound sleep."

"On the way back tomorrow, I know a cave where there are undisturbed quartz crystals!" Ki' mentioned. I gave him a thumbs-up. He remembered my DNA contained rockhound-genes.

We left Surprise Valley passing Owl Bridge, a red sandstone arch next to a stand of cottonwoods guarding a patch of lavender sego lilies, confirmation we were on the trail. Once we got back up, close to a mile high, we rode rectangular-rock outcroppings to the north and enjoyed the expansive view.

Oak Creek Canyon came next. We rode into it and let the horses drink in the first creek we came to. Ki' and I dismounted and had lunch: venison jerky with crackers and an apple. Then Ki' rewarded each of the horses with an apple. Amazingly, we had not met or seen anybody else all day—true freedom.

Moving on, we wound and bumped our way on switchbacks heading out of Oak Creek Canyon. This is when we encountered a serious adversity: There was a gaping, five and a half foot section of the trail missing on a steep switchback next to a rock cliff. We immediately dismounted.

It appeared that a large rock had fallen from the cliff onto a flat rock that had been part of the trail. There were other small rocks on each side of the "hole" and a seven hundred foot drop to the canyon floor on our left. The horses were disturbed, there was no room for them to easily turn around. I guided Madonna as she slowly backed up, giving Tonto more room. Ki' kicked small stones down into the canyon, smoothing out the trail next to the hole.

"The horses can jump it, but not with us on them. Can you jump it?" he asked me.

"Maybe with a running start in my younger days. Hard to do confined by the switchback. Is there an alternate trail?"

"This is the only trail!" Ki' frowned. We both went over to the hole and took a closer look.

"If I can get a rope across that hole and tied to a rock, can you pull yourself over on it?" Ki' asked me. I looked down; the hole didn't go through to the canyon floor. I could see the two rocks that had collided about twelve feet down in a cavern below. The one that fell from above was made of Kayenta sandstone, the other looked to be made of the softer Navajo sandstone.

"I can do it! How do you intend to get a rope across and tie it?"

Ki' had disappeared, he was unpacking a rope from one of Tonto's saddle bags. He tied it to Tonto's saddle and made a double loop with the other end under his arms. Madonna watched him as he took Tonto by the reins to the edge of the hole. The horse looked down. Ki' gave him time to study the predicament. The animal whinnied. Ki' then explained something to both horses in Navajo.

''Stan, when I get across, I'll throw this loop back to you. Tie it to a rock. I'll tie the other end over here."

"Understood!" I said loudly. Strangely, I was more worried for the horses than myself. I was thinking I could climb down into the cavern using rocks on the cliff side and then back up the other side with the rope tied to me for security. Ki' could do the same. But what about the horses? He had to jump across with Tonto first.

Without further discussion or delay, Ki' backed Tonto up as far as possible, opened the rope to slightly more than twice Tonto's length, then slapped the horse's ass hard with the rope while yelling "DEEYA HAH!" (jump quickly in Navajo). Tonto bounded over the hole, all four feet in the air like an African Impala, and made it across with a foot to spare. Ki' had waited for the slack in the rope to tighten, then—timing it perfectly—he jumped over the front edge of the hole just before Tonto landed on the other side. Ki' hit waist-high on the far edge, suspended by the rope, he kept his feet moving as Tonto continued forward, pulling him up.

"Are you okay?" I yelled. He gave me the okay sign.

I studied the new situation and could see Madonna was nervous, so I spoke to her confidently. She was bigger than Tonto and couldn't jump quite as well.

"I'll move Tonto. Madonna's next; back her up." Ki' shouted. I knew then I could make it across alone with the rope once it was tied to a rock. I patted her on the head and backed away. Ki' whistled loudly, and instantly she was over without an inch to spare, having kicked a scary amount of dirt and rock down into the canyon.

"She just made it!" Ki' shouted. I could see that; it made me shudder.

After reconsidering again, I decided doubling rope and tying it extra tight on each side of the hole would be best for a hand-over-hand crossing. I shouted this to Ki', who agreed. Fortunately the rope was a thirty foot rodeo rope; I doubled it and in short order Ki' and I had it tied tight, spanning the hole close to the cliff's wall. I wished I'd had my nylon skiing gloves to grip with. They were back in Minnesota. I dropped that thought, grabbed as close to the center of the double rope as I could reach from a kneeling position, waited a few seconds for Ki' to get in position—then pushed off, staying close to the wall. Five hand-over-hands with my feet repelling off the wall of the cliff got me to where Ki' could grab my vest and pull me up. Whew!

Leaving Oak Creek Canyon, we rode on, eventually joining up with the south trail where both trails ran together for two and a half miles leading to Forbidding Canyon. I was excited, but there were still challenges ahead. We walked the horses to rest them.

The north trail to Forbidding Canyon is thirty-one miles round trip from Navajo Mountain; it's the "old way" trail we'd been on. The canyon has two arms, both of which have river basins that drain into Lake Powell. This area, only accessible by water since the 1960s, was undiscovered by white men up until 1909. The eastern arm, also called Bridge Canyon, is the main reason I took this horse expedition. It's home to Rainbow Bridge: a perfect rainbow arch, two-hundred-ninety feet high, forty-two feet thick, and thirty-three feet wide. It's in a different league from all the other arches in Utah. Formed first as a wall during the Jurassic period from highly compressed sandstone, it then experienced various periods of flowing water, from receding seas to redirected snow-melt from Navajo Mountain. After the last ice age, the developing bridge was confronted by powerful, swirling eddies of water which finished the job of carving its perfect arch. Called *Nonnezoshe* by the Navajos--a rainbow turned to stone—Rainbow Bridge is a sacred religious site for them. The area was dedicated a National Monument by President Howard Taft in 1910, less than one year after John Wetherill's party discovered it. President Teddy Roosevelt and Zane Grey were among the first to visit Rainbow Bridge by horseback.

Here's how Zane Grey described the thoughts Shefford had when seeing Rainbow Bridge for the first time in the 1915 novel, *The Rainbow Trail:*

"He had a strange, mystic perception of this rosy-hued stupendous arch of stone, as if in a former life it had been a goal he could not reach. This wonder of nature, though all satisfying, all fulfilling to his artist's soul, could not be a resting place for him, a destination where something awaited him, a height he must scale to find peace, the end of his strife. But it seemed all of these. He could not understand his perception or emotion. Still, here at last, apparently, was the rainbow of his boyish dreams and of his manhood—a rainbow magnified even beyond those dreams, no longer transparent and ethereal, but solidified, a thing of ages, sweeping up majestically from the red walls, its iris-hued arch against the blue sky."

Forbidding Canyon's east arm becomes Bridge Canyon with Bridge Creek flowing through it, coming to Rainbow Bridge, then ultimately Lake Powell. In high water years, the water running below Rainbow Bridge can be fifty feet deep. Aztec Creek flows through the west arm of Forbidding Canyon and into Lake Powell, but completely misses Rainbow Bridge. Some early explorers took the western arm by mistake.

Bridge Canyon besieges you with eight-hundred-foot red walls streaked with stains of desert varnish caused by bacterial action: black from manganese oxide, yellow-orange from iron oxide. On some walls the mineral streaks are perfectly parallel. It took us two hours to get through it—half riding, half-walking—including hopping boulders in the creek. We came into Echo Camp and were pleased it was empty. There was no camping allowed in the National Monument, and this was as close as we could get to make camp. A large, dead cottonwood tree, a rock fireplace, several sets of rusty bed frames, flat rocks to sit on, and an old, steel drum stove greeted us. We set up the tent next to a patch of scrub oak and organized some firewood for later. We were close to *Nonnezoshe,* and wanted to see it before sunset.

I was not disappointed. It was even more majestic than I had imagined. Nobody else was here. Close to the leg of the arch on our side of the river, we dismounted and walked up to where we were almost under it. Ki' said a Navajo prayer, which allowed us to walk under the Rainbow. The silence was magical; I could tell even Madonna sensed it.

I felt the energy of the place. Mesmerized until Ki' pointed to a flat rock ridge next to the river.

"What's over there?" I asked.

"That's where Teddy Roosevelt took a pee."

"You're kidding me! That's sacrilegious," I blurted. "You told me he'd been standing on a cliff?"

"That was another time, he had to go more than once." Ki' grinned.

"You're full of donkey dung!" I was trying not to laugh.

Ki' slapped me on the back, then mounted Tonto. "Let's water the horses and go make dinner. We'll come back."

Echo Camp was not so bad after all. Two of the bed frames would work. Ki' put a Mylar blanket over one, then doubled his wool blanket on top. I pushed a frame into the tent and covered it with my sleeping mat and bag.

Ki' built a roaring fire; there were lots of scattered cottonwood branches to use. I cleaned up an old kettle that a previous guest had left and filled it with river water, then boiled it for fifteen minutes.

"So long boiling?" Ki' asked

"We have to kill all the germs, particularly the tough ones."

"I studied history at Arizona State—never had much science," Ki' lamented.

"I'd say you know plenty of nature science. You're a smart guy Ki' Somma; never sell yourself short." I cut open two bags (four servings) of Mountain House freeze- dried Spaghetti in meat sauce. Pouring them into the kettle of boiling water, I added several pinches of oregano and stirred briskly for eight minutes. Ki' had two paper plates and plastic forks ready, plus a small bottle of Navajo hot sauce. After several shakes of hot sauce, dinner looked and smelled good enough to eat. We drank water from the river that I'd purified through a 0.2 micron filter. I put two shots of Chambord in each of our canteens with the water—for flavor.

"Bon appétit," I proclaimed.

After dinner we walked with the horses who found edible grass along the creek.

Nonnezoshe, "the rainbow turned to stone," was stunning in the moonlight. Chirping frogs violated the silence we'd had earlier, but that was okay. They were part of this picture. We stretched out on the grass under the north leg of the bridge and looked up. The stars were out in strength except for a wide, opaque black band above us—the span of Rainbow Bridge, of course.

"It divides the universe in half, one half is east of the bridge's span and the other half is west," Ki' revealed.

"I'll drink to that!" I took a drink from my canteen. "Don't let me drink with the flies, Ki'."

He opened his canteen and took a drink. "I know what that means. Last year I guided a tall Australian man. He explained it meant it's not wise to drink alone."

"Then have another for his sake," I told him. I did the same.

"When did you first hear about *Nonnezoshe*?" Ki' sipped from his canteen.

"When I was ten years old, I saw Tom Mix on a comic book cover. He was on his horse under a giant stone arch, somewhere out west. Cowboy Tom and his horse were dwarfed by the arch—Rainbow Bridge. As I grew up, I couldn't get that picture out of my head—knowing that someday I would need to confirm it was real."

"It's a very special place for us Navajos. *Achak* likes to come here—he can feel the Rainbow's spirit. Old rocks have spirit."

"He's your father, right?"

"Correct. He runs the family gas station in Kayenta." Ki' looked tired.

"What's the plan for tomorrow?" I could tell I'd have no problem sleeping.

"We'll take the south trail back—don't want to encounter that hole in the trail at Oak Creek Canyon again."

"Don't you want your rope back?"

"My brother will get it next week; I assure you it'll be there. I'll have my licensed nephew tell the Canyon Institute and Park Service where the trail is damaged." Ki had trouble keeping his eyes open.

It was 9:00 p.m. My knees had experienced a substantial workout—even more than after ten hours of casting for muskies in Canada. The trail incident in Oak Creek Canyon seemed like days ago. The horses were loose down by the creek eating grass. Madonna was in charge. It felt good to stretch out. I recall mumbling, "Let's stay here under the Rainbow for a few more minutes."

I jerked suddenly and woke up. Madonna was nudging me with her head. Ki' was snoring. "What the bejesus time is it?" I activated my Casio. "My god, it's 2:00 a.m."

I shook Ki'. Had to do it twice. He exclaimed something in Navajo, then said, "Who are you?"

"I'm Stan—your guest."

"Stan—yes, of course. Sorry, I was dreaming."

"We fell asleep, it was the Chambord."

"Oh yes, the special raspberry juice."

As moonbeams illuminated the Rainbow, we stood up. Before the short walk back to Echo Camp with the horses, I looked again at the two halves of the universe, thinking how fortunate I was to know that a divine path ran through the complexity of it all. And possibly, that the path was curved, maybe like a giant, perfect arch, like a bridge that had turned to stone.

Postscript

The south trail had its own special challenges of which Redbud Pass was the most memorable. The horses handled it well. Riding Madonna, I could touch both sides of the canyon as we moved through the narrow passage slowly. At one point we had to shovel rocks off the trail. Two years later, in 2004, a rockslide blocked the pass to horses. Hiking became the south trail's only mode of access after that.

As of 2019, an enterprising guide service provides group trekking trips on the north trail, using llamas to pack in supplies. It's a fourteen mile trek to Rainbow Bridge with four nights of camping, then a return by boat to Page, Arizona's Wahweap Marina—a fifty mile ride on Lake Powell. The easiest way to visit Rainbow Bridge is to invest a day and go by boat there and back the same day. From where the boat docks, it's an easy 1.5 mile hike to Rainbow Bridge. Engage your soul voice if you need advice.

Sydney Opera House and Manly ferry

Chapter Nine

Eleven Miles from Sydney

Comprehending the Mysterious God Particle

Sydney, Australia—December 2011

I was having breakfast on the rooftop restaurant at the Top of the Town Hotel in Kings Cross, or "The Cross" as locals call this part of Sydney. Twelve stories below on Victoria Street a marching band, more than slightly out of tune, created clamor normally absent in the morning on this end of the street.

"What's going on downstairs?" I asked the waiter, pointing outside. It sounded like a first practice session for a junior high school band—lots of discordant trumpeting and hard pounding on drums.

"The professional Sheilas are out protesting this morning." Blake smiled as he refilled my coffee.

"Really? That explains the band being out of tune," I smiled back. "What are they protesting?"

"They want better government healthcare for prostitutes."

"Does it vary from state to state in Oz?" I knew prostitution was legal in Australia, and that Aussie Medicare covered everyone.

"New South Wales has the most liberal policies for sex workers, but there is discrimination. Doctors decide on what treatment to use. Prostitutes don't always receive the best treatment. People with private supplemental insurance do."

"Hmm, didn't know that. The noise is certainly getting a lot of attention."

"It's not the noise that's getting the attention. Many of the Sheilas are topless," Blake informed me. That explained the collection of men standing at the windows, intensely observing the protest. Blake seemed surprised I didn't stand up to look.

"I have a meeting in Manly later. There'll be plenty of topless women on the beach there," I said to clear up his puzzled look.

"Manly is only eleven miles from Sydney," Blake noted. "The ferry will take you directly to it."

"I know; been there before." I took a drink of coffee. "The meeting in Manly will delve into how life can exist in our chaotic universe." Blake got that look I've seen before which said, *he must be one of those Yankee nerds who didn't know when to quit school and go to work.*

The meeting would be on a beach where topless bathing was optional. But I'd be with a group of graduate eggheads who studied advanced math and physics. We would review a recent discovery in quantum physics. Who knows how they might be dressed? Anything from full lab coats to string bikinis. Their professor told me his students would appreciate spending a December Monday on a favorite Aussie beach rather than sequestered in a hot classroom surrounded by blackboards. The protest was already heating things up outside the hotel.

My plan for the morning was simple: After breakfast I'd walk William Street west about eight blocks to College Street, then head north to the Royal Botanic Gardens; about a half mile.

The huge gardens are a favorite go-to place for me in Sydney. It's easy to find a park bench around the main pond with a view of the Opera House and the Harbour Bridge in the distance. I stretched my legs out on such a bench and focused my binoculars on some birds.

A plethora of them inhabited the pond. A couple dozen Australian White Ibis were undisturbed by my presence. These large birds have obsidian-black heads, curved bills, and long legs attached to white-feathered bodies; they waded in shallow water in search of bugs and minnows—their daily game plan. I'd seen versions of them in North Africa. Although unpopular in backyards and swimming pools, they complemented the botanic gardens' ecology perfectly. One large White Faced Heron kept an eye on the ibises—his unwelcome competitors.

Sulfur-crested cockatoos and rainbow lorikeets were easy to spot in the broad shade trees around the pond. Tourists largely ignored the posted requests to not feed the birds. If you're a birdwatcher and you notice tourists around a large deciduous tree in Sydney, focus your binoculars up in the tree.

When I got up to walk around, my movement startled the heron, who awkwardly took flight. The ibises, with no cares beyond finding food, didn't react to either me or the heron leaving.

There are areas in these seventy-five acres of gardens that deserve special attention—even on revisits. The Palm Grove is a proven place to spot fruit bats or "flying foxes" that call the gardens home—they're huge, hairy, and hang out in some of the 140 different palm species that grow here—look up high.

After almost getting anointed by guano from heaven by several flying foxes flying, I decided to find a spot where I wouldn't be so distracted by nature, but still in the park—I had some important studying to do.

I found a bench by a cactus garden, away from birds and giant bats. I was able to mostly ignore several lizards catching flies, except for noticing something unusual: black flies were crawling on the lizards' backs, easily becoming sitting ducks to be nabbed and eaten. One lizard had three flies lined up on its back—big ones—when a second lizard close by slurped up the fly in the middle. The other two flies simply ignored this and didn't fly away! I wondered how Charles Darwin would have interpreted this . . ..

I disregarded any further entomological incongruence and redirected my attention to seriously studying two scientific reports from CERN (Fermilab) in Europe. I had to read these thoroughly in preparation for the discussion in Manly. Since I'm a biochemist, I needed to sharpen up on elementary particle physics before attempting to challenge young physicists and mathematicians. CERN is the French acronym for what translates in English to: European Organization for Nuclear Research. Founded in 1952 with the mandate to become a world-class physics research organization, part of its scientific mission was to build particle-smashing machines that could reveal what the universe is made of. In other words, discover the reasons how—and maybe why—everything exists, including us.

After "work" (studying the documents) I headed to "The Rocks" area by the harbor, within sight of the ferry docks. It was lunch time and the beginning of a sunny summer *arvo* (afternoon) in Sydney.

About a mile's walk from the gardens, I was back in the 1800s: George Street in The Rocks. Lots of bars, restaurants and shops with a vast array of merchandise lined the streets. One claimed to sell authentic Aboriginal art and hunting gear—like Boomerangs and *waddies* (sharp sticks). Two protestors outside that store held a sign: "Ban *Waddies* and Boomerangs!" Australia already had strict gun control.

"Those *bogans* need a job," a stately-dressed elderly man with curly white hair proclaimed as he walked by the protestors. He reminded me of Mark Twain.

Another man—younger and dressed for the beach—informed the elderly gentleman, "Those bloody *bogans* work for the store. It's marketing warfare—claim something will be banned and it sells like *pikelets*." I had to smile; this was surely the Land of Oz (*pikelets* are Aussie pancakes).

The cove begins at the southern end of the Harbour Bridge with the Opera House across the water on Bennelong Point. It's a rush of activity with sailboats and motorboats making way for the ferries, while out further, wave runners and jet skis take on the challenge of choppy water. Eleven miles west on the Pacific Ocean—thirty minutes by ferry—is the beach city of Manly, my destination later in the afternoon.

I found my favorite Rocks restaurant with outdoor dining in full view of the Opera House which, in the midday sun, showed off its exceptional architecture of white, oval, overlapping, geometric "sails"—unique by any standard.

I started with a pint of Victoria Bitter (VB) then ordered fifteen freshly-shucked Sydney rock oysters. "Can I get chopsticks and three whole lemons cut in half with the order?" I asked Julie. Her name was on a cricket cap she wore over her blonde ponytail.

"Maybe," she said. "We might have to charge extra for three of them."

"No worries, I expected that. The lemons are important."

"How about some fresh baked sourdough bread to go along with your oysters and lemons?"

"Sounds like a ripper! Let's do it." She smiled and went to place the order. I was always careful when using Aussie expressions to *not* fake an accent—I just say them in my Midwest American accent. This usually gets a smile rather than a sarcastic laugh, at least from the women.

Sydney rock oysters, indigenous to Australia's east coast, are the best oysters in the world. They're the only ones I eat raw. Oyster connoisseurs claim their taste has a level of complexity not found anywhere else. Once I get them pH-adjusted with plenty of lemon juice, I touch the ends of my chopsticks in Cholula hot sauce (brought it with me!) then pick an oyster out of its shell and eat it, getting just the right amount of sauce on the oyster! The fresh, cold sweetness of the oyster with its tinge of sea veggie-ness melds with the lemon and Cholula, creating a flavor masterpiece. When chased down with fifty milliliters of cold VB; it might approximate dining in Nirvana.

When Julie brought out the stainless steel tray of iced oysters on their half shells with six half lemons, it got the attention of diners at two nearby tables. I proceeded to rain down lemon juice on the mollusks, energetically compressing the lemons to get every last drop!

"Need to get the pH close to 2.0 for ten minutes to kill any toxic microbes," I declared [1]. Everyone just stared. One woman nodded repeatedly.

"Do all you Yanks do this? We may need to order more lemons," Julie announced, as she looked around to see the other customers gaping at me squeezing the lemons. I winked at her and shook my head no, then noticed her tee shirt: "The Top End Rules." *Cute,* I thought.

"Are you from the NT?" I asked her.

1 Raw oysters can be contaminated with viruses and bacteria, like the Noro virus or Vibrio bacteria. Lots of lemon juice lowers the pH to 2.0 and kills these bad bugs, rendering the oysters safe to eat. In addition, it's always smart to take a double dose of Lactobacillus probiotics after eating oysters.

"Yup, born and raised in Darwin." The nickname for the NT, Australia's Northern Territory, is "The Top End".

"Do you know Darwin?" She asked me.

I was still squeezing lemons. "Quite well! Love to fish for big barramundi in the Roper River."

"So does my older brother; he's a guide." That got us talking about "Barra" fishing and got me a telephone number—her brother's.

When I first became aware of the God Particle in 2000 its existence hadn't been proven yet, but theoretically it was on firm ground. All of its unique characteristics had been predicted—a massive, unstable, elementary particle with no spin, no electric charge, that confers mass to the universe. An elusive particle that allows all things to exist as they do—including people. If found, it would be unlike any other subatomic particle and would complete the Standard Model of Elementary Particles—further confirming quantum theory.

Finding the God Particle was vital to proving how atoms could exist in the universe. An advanced super-collider would be required to find (or dismiss) The God Particle, a.k.a. the Higgs Boson, as physicists call it. The collider would operate at energies up to 13.5 TeV (trillion electron volts), equivalent to a freight train running downhill at full speed. It would be called the Large Hadron Collider (LHC).

Construction of the LHC started in Switzerland 1996. Over thirteen billion dollars and fourteen years later, in March 2010, it was ready to run. By early 2011 there were tantalizing hints that a mystery particle of the predicted mass for the Higgs Boson had been discovered, but greater statistical evidence (more particle-collision events) was required. Excitement started to build at CERN.

The LHC is the most complicated machine ever built. It's enclosed in a circular tunnel twenty-seven kilometers in circumference, surrounded by giant, cryogenically-cooled niobium-titanium magnets. The whole works is 150 meters underground between France and Switzerland. It can smash a hundred trillion protons into each other at once at 99.999999% the speed of light. The collision events (particle decays) are recorded as colored starbursts on particle detector screens that reveal the inner complexity of the colliding protons.

For those readers who would appreciate an update on a few basic facts from high school chemistry and physics before I move on with the story, see footnote number five [2].

When the LHC blasts trillions of protons into each other, it's using hydrogen atoms stripped of electrons—a source of protons. What's revealed on the LHC's particle detector screens are the tracks of elementary, subatomic particles either released or formed during the collisions. They comprise the Standard Model which includes quarks, gluons, Z and W bosons, photons, muons, neutrinos and—subject to verification—The God Particle, a.k.a. the Higgs Boson.

To better visualize this, think of protons as small piñatas at a Christmas party. The proton particles are represented by small donkeys made of *papier mâché* and filled with candy. Kids hit the piñatas with sticks, trying to break them open. A low-energy hit releases only a few pieces of "candy" (elementary particles), if any—like you might get from knocking an ear off a donkey. A medium-energy hit might knock a leg off and release a handful of candy—a greater variety of different types and flavors, whereas a super high-energy hit breaks the piñata wide open, all candy types (elementary particles) are released, and maybe even some new ones form. The LHC was designed to provide super-high-energy hits. *To know everything that's inside something, you must break it completely apart.*

---

2 Atoms are made of protons and neutrons that reside in the nucleus and electrons that reside outside the nucleus in spaces called orbitals. The protons and neutrons are composite particles and, unlike electrons, are not elementary because they can be further broken down. The simplest atom is the hydrogen atom (also the first element in the periodic table): one proton forms the nucleus and one electron occupies an s-orbital.

The number of protons determine an element's atomic number and identity: Sodium (Na) has eleven protons, Iron (Fe) has twenty-six, lead (Pb), eighty-two. There are ninety-two naturally occurring elements, uranium is the last with ninety- two protons. Since all protons have a charge of +1, atoms must contain an equal number of electrons, each with a charge of -1, to maintain electrical neutrality.

Neutrons also reside in an element's nucleus to provide stability—they have zero charge. Hydrogen is the only stable element that has no neutrons. The number of neutrons added to the number of protons yields the atomic weight(s) of an element. Isotopes of an element have different numbers of neutrons but always the same number of protons.

Manly Beach, same day

Graham O'Brien drove onto the beach to unload firewood. He's a superb organizer and my business partner in South Africa.

I talked with two of the graduate students while six others were slinging Frisbees. A stray dog played along with them, hoping to retrieve a cast that mistakenly went into the ocean.

"You sure we can build a fire on the beach?" I looked at Graham.

"Not sure, but optimistic! The professor said it would be okay down on the north end, as long as we clean up after and don't get rowdy."

"How rowdy will you guys get?" I asked the two female grad students I'd been talking to. They chuckled at being called guys.

"Not very. It's an alcohol-free zone!" Joni lamented, straightening her University tee shirt. Her Peter Pan earrings and silver-speckled toenails glistened in the afternoon sun.

"That's correct, no amber fluid is allowed!" Helen nodded firmly, sending her eyeglasses off kilter. She wore a tan pantsuit with her hair short and combed back. With two ballpoint pens in her breast pocket, she reminded me of my high school English teacher.

Both attractive women in their mid-twenties, Joni and Helen were studying quantum computing in graduate school. Helen was not the least ostentatious—no tinsel.

Of the six male students throwing Frisbees, four were physics grad students and two were advanced math majors—only one seemed to have any athletic credentials; "Nathan James" was stitched on his hat.

"If we talk about The God Particle tonight, we'll need to be sober," Helen asserted. I agreed. "No liquid-variety spirits at this party."

"Look!" Graham pointed over my shoulder at the ocean. Two men on long surfboards were surfing in parallel, one twisting slightly right and the other slightly left—perfectly in sync on a large, breaking swell.

"That's the surfing Manly beach is famous for," Joni declared. "And gnarly (awesome) surfers!"

"Supersymmetry," Helen declared, pointing at the surfers. "You sure there's just one type of God Particle on the agenda tonight?" Supersymmetry is a transformative, debatable concept in particle physics.

I laughed, "One so far, but you're welcome to redirect the discussion, if you know any God Particles that spin." Supersymmetry requires elementary particles to have super-partners with opposite spin, which opens the door to "new physics" beyond the Standard Model of four forces and a handful of elementary particles. Since the God Particle has no spin, finding a super-partner is unlikely.

She gave me a sly smile, "That may be too off-the-charts for the math majors."

"Don't be so sure." Joni looked at Helen. "Nate says he can integrate fractals in multiple dimensions." Nate was the tall, athletic math major who had begun purposely throwing the Frisbee into the ocean for the dog to fetch. An alarm went off in my head! *It didn't take long for the conversation to deepen.*

"Look who's here!" Graham yelled out as he fanned the fire he had started. Professor Johan Hansen had just walked onto the beach, smiling and waving to us—he gestured his approval of the fire. Johan, born and raised in Denmark, had earned his PhD in quantum physics at Stanford. Johan, Graham, and I had spent a marvelous ten days together in the South African Bush the previous year—we got to know each other well.

"G' day Professor," I exclaimed as I walked over to him and shook his hand. He was the students' fearless leader. "It's great you could make it!"

"I rescheduled the administrative meeting, didn't want to miss a discussion on The God Particle with you, Graham, and my gang of geeks."

"Helen's already talking supersymmetry!" I shook my head.

"We'll have to keep the reins tight on that one!" he said loudly enough for her to hear. The Frisbee event came to a quick halt when the six young men saw Doctor Hansen. Everyone gathered around the fire and began drinking iced rooibos tea thanks to Graham. He had prepared five gallons of the caffeine-free red tea from South Africa and had it iced up in an esky (cooler).

The professor raised his paper cup and proposed a toast to Graham and me, "*Skoll!*" Everyone drank the special tea.

"This tea is full of antioxidants and will keep you two math majors from freaking out tonight." Helen smirked; she was clearly not a fan of Nate's dogmatic approach to science.

"Helen, don't start any fights you can't finish," Johan advised. He knew she really admired Nate but viewed his lifeguard image as too far off the mark for a graduate mathematician. He was too much of a dude for her.

It was a perfect 24° Celsius late *arvo* in southeast Australia. The famed Manly beach stretched out in front of us for 1.4 kilometers. A surfer's paradise. The tide came in with a gentle breeze, enough to sway the tops of the tall Norfolk Island Pines that lined the road behind the beach.

Graham had organized this fireside chat at Doctor Hansen's suggestion. He was eager for me to share with them an update on CERN (Fermilab) from my last trip to Geneva, and the students were particularly keen to hear what progress had been made with confirming and understanding The God Particle.

By sunset Graham had the fire going strong. The students were primed and excited to have a serious chat, as Professor Hansen preferred to call it. It was not usual to have a scientist well outside their field of expertise speak to them. I have a master's degree in biochemistry, not particle physics or quantum math; but, always fascinated by these subjects, I'd learned much from studying and conferring with physicists.

"Thank you lads and lassies for coming tonight!" Graham started out with more Irish brogue than normal. "I'm over the moon that I could get Mr. Stanley Randolf to find time in his world travels to meet tonight. He has an early flight to EnZed in the morning so we can't keep him up too late.

"Let's agree to keep everything informal and ask any questions or make any comments that come to mind. The subject of The God Particle, a.k.a. the Higgs Boson, is a strange mystery, likely woven into even stranger mysteries. As you know, it has to do with how atoms formed after the Big Bang—all the odds were against this happening.

The universe should not contain any life today. Life demands atoms. The odds were less than a hundred billion to one against atoms forming from a lifeless soup of quarks, gluons, and electrons! Explaining anything more is outside my job description, I'll let Stanley take it from here. . ..“

I thanked Graham, Professor Hansen and the eight students for coming, then reinforced what Graham said about asking questions or making other comments—all would be welcome. "The tour group at CERN had no problem firing dozens of questions at Nobel scientists, so feel free to attack me." Everyone listened quietly.

"Let's start by taking a hypothetical stroll through the Big Bang and the early universe to set the stage for The God Particle and why it deserves that name. I've divided it into eight time periods or chapters—so don't fall asleep. Johan told me aspects would be on your next exam." That got a few chuckles.

"Chapter One: The universe began as a singularity, significantly smaller than the head of a pin, thirteen billion and eight hundred million years ago." I looked up and moved my hand across the southern hemisphere's stars which had started to show, stopping at the Southern Cross. "Yes, the potential for all of everything was somehow inside that head of a pin. At last count this translates to over one hundred billion galaxies, each hosting a hundred billion stars or more. It must have been a heavy pinhead! Or was it? One of the scientists at CERN reminded me that the 'potential for something' doesn't require it to have weight or mass. How much weight does a thought have?

"Chapter Two: Just a few milliseconds after the pinhead decided to explode, the temperature reached ten billion degrees Kelvin, and gamma ray photons—with unimaginably high frequencies—streaked out. Energized by trillions of electron volts, they reacted with each other and formed the first electrons. This didn't take very long—maybe a few seconds.

"Chapter Three: When the universe was a minute old, it had become a primordial soup of quarks, gluons, and electrons. All massless, all traveling at the speed of light. Hard to imagine even when sober." I smiled at Helen.

"Chapter Four: At three minutes of age, the universe had cooled enough to allow gluons to bind quarks together forming protons and neutrons and then, began binding these into nuclei. This ability to bind positively-charged particles—which normally repel each other—is manifested by the strongest force in the universe. It was given a clever name by physicists. As you all know, it's called the *Strong Force*." That got the two math majors to laugh.

I toasted their laughter with a sip of tea. "It was the beginning of nucleosynthesis: Hydrogen, helium, lithium and beryllium nuclei were produced. *Not atoms, only nuclei*! These nuclei had positive charges and about ninety-five percent of their mass was due to binding by the strong force. Some particles got mass, others didn't. Electrons had no mass! You could say the universe was partially materialized and now had some weight.

"Are you all okay with the length of the chapters?" Johan and the eight grad students smiled and nodded. "This was a dangerous time, so we're strolling fast—very fast.

"Chapter five: Around two hundred and fifty thousand years later, the universe cooled enough to potentially allow atoms to form from the nuclei of those first four elements. But it didn't happen. The universe had changed into to a primordial plasma—hot ionized gas—and was expanding rapidly. Particle interaction could not effectively occur. And when interaction attempted to occur, massless electrons could not be captured and held in orbitals surrounding nuclei. The odds of a massless electron being captured by a nucleus and forming a stable neutral atom was essentially zero."

I paused and took a drink of the red tea. Good thing it was alcohol free. Joni was petting the dog. It appeared to have been listening intently as well.

"Chapter Six: At some time later, but not much later, the Higgs Field turned on! It was like someone had poured an ocean of cosmic molasses into the plasma--reducing the degrees of freedom available at the subatomic level. Particles that slowed down gained mass as they passed through this viscous Higgs Field. With help from Higgs Bosons, particles that slowed the most gained the most and, conversely, those that slowed the least gained the least. Higgs Bosons, created from perturbations in the Higgs Field, made it all happen. They finished the process of giving mass to the universe. Protons became one thousand eight hundred times heavier than electrons—but, nevertheless, electrons now had mass! *And atoms could form*!"

Note to readers: Atoms are required for life to exist. We're all made from atoms of the 92 naturally occurring elements—eleven elements account for 99.85% of the human body.

"Chapter Seven: Why did the Higgs Field turn on when it did? It was a critical time for it to do so! Had it happened much later, the mathematics of inflation indicates that the Universe would have expanded to such a size that the density of particles would have been too low for effective interaction. Had it turned on too soon, not enough nuclei would have existed. This event alone justifies the Higgs particle to be called the 'God Particle' and the Higgs Field, the 'God Field.'"

"Chapter Eight: First generation stars began to form approximately one million years after the Big Bang. As this process advanced, galaxies containing different types of stars formed and, from them, nuclear fusion reactions, fast rotating binary stars, and super nova explosions produced the other 88 elements. At the ripe old age of nine billion years, the universe gave birth to our sun and solar system."

I paused to look around; nobody had fallen asleep. Johan was grinning. The surf and the crackle of the fire were the only sounds. The students were bright-eyed and alert. Presumably, they'd heard descriptions of the Big Bang before—probably more than once—but they seemed to appreciate an updated version that included The God Particle's astonishing appearance.

"Intermission" I said loudly. "Exercise!" Everyone got up and kicked sand, including the professor, then they held hands and danced around the fire singing *Waltzing Matilda.* I played the ageless tune on my harmonica. Everyone, including the stray dog, had fun while Graham refilled their teacups.

Nate spoke up after the break. "So perturbations of the Higgs Field occur when trillions of protons collide at near-light speed, creating Higgs Bosons that quickly transfer mass to other particles?"

"Correct. So maybe we should be talking about God's Energy Field as the elephant in the room," I proclaimed, looking around for some reaction.

Nate quickly responded, "*Deadset (unquestionable).* It's the Higgs Field that holds massive energy potential for the future. The Higgs particles are just delivery boys for tiny bits of mass—they disappear after making a split-second delivery."

Helen was standing in front of the fire with the ocean to her back. She appeared surprised by Nate's cogent comments. "Is it true that the Higgs Boson has a lifespan of a septillionth of a second?" Helen asked me.

"That's what they've found at CERN, 10E∧-22 seconds: Ten to the minus twenty- second power! It gives new meaning to a split second."

"Faster than a lizard can drink," Graham added some Aussie humor.

"How many collision events does it take to get a Higgs particle on the LHC?" Professor Hansen asked.

"It takes one to two billion proton collisions to produce one Higgs Boson particle!" I replied. "That's about two per second with the LHC running flat out. They expect to improve on this in the future with higher energy collisions."

"Not a very efficient process!" Nate noted.

"What statistics do they have on this?" Johan Hansen hand-combed his white hair. The tide was coming in with a stronger breeze.

"It's early in the experimental investigation. They're close to three sigma right now but require five sigma before announcing a discovery." I flipped through one of the reports, double checking what I'd said: "Three sigma equals a ninety-nine-point-eight-seven percent chance the measurements are real—not noise. Five sigma equals a three-and-a-half million-to-one odds the measurements are real. CERN insists on five sigma to quell any doubts!"

"How much longer will it take to get that?" Graham stoked the fire.

"Good question! They didn't say. As a research scientist my answer would be: It won't take one minute longer than it takes." That got a few laughs.

"Actually, it'll require numerous additional 'proton-collision runs'—a year is my guess." I was still flipping through the report. Nate stood next to me reading over my shoulder, so I handed him the report. I'd read it three times.

"It gets damn hot inside that LHC," Nate barked, surprising the dog who barked back.

"It must take tons of liquid nitrogen to keep twenty-seven kilometers of thirty-five ton magnets cold," the professor remarked.

"During operation, the LHC reaches an internal temperature one hundred thousand times hotter than the sun; tremendous cooling is required. The young French physicist doing our tour simply said '*énorme quantité*,'" I explained. "French scientists have a special way of making you feel satisfied with an answer that doesn't tell you anything." That got more students to laugh.

"Here it is!" Nate found it in the report. "Ten thousand, eight-hundred tons of liquid nitrogen at negative one hundred ninety-six degrees Celsius is required for cooling the LHC, plus one hundred twenty tons of liquid helium at negative two hundred seventy degrees Celsius. This cools the magnets to one-point-nine degrees Kelvin. Colder than outer space!" Nate added.

"Zero Kelvin is absolute zero—the coldest anything can *never* get," Helen commented. The third law of thermodynamics asserts that absolute zero can never be reached. Helen was staring straight faced at Nate. Nate—Nathan James—had long stopped making passes at Helen. Nevertheless, he smiled and winked at her when she stared at him.

"That's colder than a well digger's arse in Minnesota in January!" Graham declared, generating more student laughter. They knew my company was based in Minneapolis. We took a short break to move around again and kick sand off our legs; everyone except Helen and the professor wore shorts. Joni had a tin of homemade cookies that went well with the tea. Nate gave one to the dog which was still hanging around. That got a dismissing look from Helen.

Professor Hansen asked Joni a question as I stoked the fire: "What would Albert Einstein say about the Higgs Field if he were still with us?"

Joni thought for a moment. "Doesn't the Higgs Field resemble what he called the cosmological constant?" Joni asked.

"Yes," the professor answered. Go on. . .."

"I believe he would feel relieved. And pleased. He called his prediction of a universal energy field—the cosmological constant—his biggest mistake. Actually, it's turning out to be one of his most brilliant predictions. There was no LHC to prove him right or wrong when he died in 1955."

"Very good!" Johan smiled at Joni.

"How much 'energy equivalent mass' does the Higgs Field have?" Scotty, Nate's fellow math student, asked—his first words of the evening.

"There is debate on that. One of the esteemed physicists at CERN claims it's 10E∧46 joules per cubic meter; ten to the forty-sixth power. Think of what that might represent for the space inside my teacup," I suggested.

"*Caramba*!" Doctor Hansen mumbled.

"And he doesn't mean the town in Queensland," Graham piped up while Scotty worked his laptop at full speed.

"Let us know what you come up with, Scotty," Professor Hansen scratched is head after hearing the potential energy in joules. A hundred-watt light bulb uses one hundred joules per second; ten to the forty-sixth power joules is unthinkable.

It didn't take Scotty long: "If you solve $E=mc^2$ for mass, $m=E/c^2$, it will require 10E∧26 Kg of mass—all converted to energy—to equal the Higgs Field in the space inside a two hundred fifty milliliter teacup."

Note to the reader: $E=mc^2$ translates to, *Energy equals mass (weight in Kg) times the speed of light squared. So, mass equals energy divided by the speed of light squared.* Einstein's most famous equation.

"Wow!" Nate blurted. "What does the bloody earth weigh?" Scotty hit some more keys on his laptop: "It weighs 6 x 10E∧24 kg!"

Helen calculated in her head. "Let's round that off to sixteen and a half earth masses. Black hole, here we come!" She half-smiled at Nate.

Note to the reader: This means that if 16.5 Earths were converted completely to "Higgs Field" energy, that amount is what would occupy the space inside a teacup. Think about how much fits in Yankee Stadium, or the state of Texas. This is no joke. You can see why Einstein called the cosmological constant his biggest mistake, and why the Higgs Field should be called the *God Field*—presuming only God can do the impossible.

We all took a long drink of the Rooibos tea, still ice-cold. "The aluminum eskys in Oz are built better than the plastic coolers in the U.S.," Graham announced as he raised his cup.

Then the professor spoke up: "It's my understanding that the Higgs Boson is quite massive at one hundred twenty-five GeV—billion electron volts—compared to the proton's zero point nine four

GeV." Mass is also reported in energy equivalents like electron volts.

"The Higgs Boson is a little over one hundred thirty-three times heavier than the proton," Scotty instantly reported with a serious face. Joni giggled. It was just simple division; she knew Scotty could run non-linear algebraic turbulence calculations.

"Interesting. I've been doing some math in my head." Nate never brought a computer; when he needed one, he borrowed somebody's. "To add some perspective, a trillion Higgs Bosons would weigh one fifth of a billionth of a gram. A two hundred fifty milliliter teacup could hold a *trillion, trillion* Higgs Bosons. Forget their instability for a moment. At a production rate of two billion per second, it would take the LHC ten million years to produce a cupful. Just an FYI."

"Wow!" was the group's unanimous response.

So far, the two male math students and the two women physics students had been contributing; the other four physics students had been silent.

Nate kept on, "I bet aliens—the ETs—know how to gently perturb the Higgs Field to get zillions of Higgs Bosons per second without spending billions of dollars! And then use that energy for something useful—rather than to simply prove it exists."

"Hold that for later, Nate. Let Stan comment on the mass of the Higgs Boson before we stray too far from Planet Earth," Professor Hansen interjected.

"Thanks, Doc. I was almost ready to head off to planet Serpo with Nate."

"Yes. The God Particle's designation is also justified by its mass: At a hundred twenty-five GeV it's in the range necessary to create a stable universe. Outside that range, instability would rule. We know now that the universe is only five percent atomic—made of atoms. Ninety-five percent is non-atomic dark energy and dark matter, which we know next to nothing about. It's a good thing God didn't wait much longer to pour on the cosmic molasses.

"And it took Him almost fourteen billion years to get only five percent of the way to the finish line!" Joni remarked while brushing sand out of her hair.

"How do you know God is a *Him*?" Helen asked Joni.

"Don't be silly and argue over pronouns, life has more important things to worry about!" Joni stuck out her tongue at Helen.

"If you decide to call God a genderless it, be sure to spell IT with two capital letters," Nate articulated, looking at Helen, who stuck out her tongue at him. Graham and Johan chuckled. I believe Nate figured he was making progress.

Nate was still stoked: "So there is no empty space in the universe! Every cubic meter of space from here through the Milky Way and beyond is exponentially packed with energy that holds itself in an impenetrable state of frozen potential—the Higgs Field—which must be perturbed to produce the "God Particle." And this requires a multi-billion dollar machine that heats up to a hundred thousand degrees hotter than the sun while smashing trillions of protons together that necessitates eleven thousand tons of cryogenic coolant to keep the machine from melting down. All for just some notable sparks on an LED screen—proof that the God Particle exists. There's got to be a way that Higgs Bosons can do something useful other than just momentarily exist!"

Another hour went by and we kept going. Nate wondered aloud how the ETs or God—it didn't matter which to him—might perturb the Higgs Field to produce exponentially more Higgs Bosons per second than what the LHC produced. He didn't believe it was necessary to smash trillions of protons together. He also believed that Higgs Field energy could be harnessed for useful purposes, something most physicists completely disagreed could be done.

"Thank you God!" Nate bellowed. "You must have first learned about your particle in cosmic undergraduate school! Permit me to suggest a post-doc assignment for you: Tell us how to perturb space such that zillions of your particles can be produced with stability and quantum coherence for everyday use. Like running that Boeing-747 that just flew over us, or carbon-dioxide sequestering machines that clean the air, or filters that remove plastic micro-particles from the ocean. . .." Nate paused and took a long drink of his iced tea—being audibly and cogently redundant was in his DNA.

Helen's ears perked up. *Maybe she'll change her mind about this tall math major?* I wondered.

"Does anybody want to add anything to Nate's comments?" Professor Hansen asked. He looked over at the four young physics-grad students who hadn't said anything yet, but had been whispering to each other.

One of the young men cleared his throat and stood up, "I'm Trevor Jones, and I will speak for the four of us physics students who have been silent. We believe the Higgs Field forms a quantum membrane that separates our three-dimensional world from the other dimensions in the universe. It's a tough energy membrane that runs from the Plank dimension, 10E∧-33 centimeters, out to the farthest galaxies—thirteen-point-three billion light years away. To solve our energy dilemma and find infinite, clean, free energy—we need to focus on an entirely new approach apart from blasting protons together. The math for doing this is quite strange."

"Trevor, Trevor, Trevor! This is the deepest thinking I've heard from you since you've been my student!" The professor declared.

"It's not just me!" He pointed to his three silent cohorts, also, young, intelligent and reticent.

"Tell us more, Trevor." I brushed sand off my legs. Graham had put another huge log from a dead Norfolk Pine on the fire. I could sense my soul voice was fully tuned in!

"The Higgs Field has a non-zero energetic resting state—this by itself makes it special. To create realistic and economical modes of perturbation with the capability of harvesting the intrinsic energy, we need to approach it from the 'other side.'"

"You mean the spiritual side?" Helen murmured.

"If that's what you want to call the fourth and fifth space dimensions—yes."

"Can you be more specific?" Professor Hansen asked Trevor, who glanced at his three cohorts. They nodded, apparently agreeing he could say more.

"It involves utilizing harmonic oscillators. Just playing the right notes on a harmonica over a bowl of hot chicken soup can potentially produce a few Higgs Bosons! You would never know it because they're so hard to detect. However, when you work from the fourth and fifth dimensions, time is flexible, which permits frequency-collation and energy-coherence operations to occur. The four of us chuckled when Mr. Randolf defined a septillionth of a second. Seconds don't exist in the fourth and fifth dimensions."

"It's no joke!" One of Trevor's cohorts yelled out. The Professor was stunned and didn't say anything. Helen, looking nervous, cleaned her glasses. Scotty was making calculations for Nate. Jodi smiled and shrugged her shoulders. The dog looked hungry and wanted a cookie.

"Go on, Trevor," I encouraged him. This was fascinating and completely off the charts.

Trevor continued: "Elementary Higgs particles operate both as waves and particles, just as photons of light do. You can think of them as frozen light. And to get their energy you have to defrost them. Harmonically-activated Higgs particles produce waves in the Higgs Field that represent modes of energy which can be harvested and used. Under the right conditions, the Higgs particles can be tuned to transfer energy instead of mass. It's all about knowing how to use $E=mc^2$. An enormous number of modes are involved; once frequency coherence is obtained, the total energy harvested will equal the sum of the frequencies cubed per-unit volume per mode. This can all be done with light—extra-dimensional, high-frequency photons. We also need to better understand how to use quantum entanglement—spooky action at a distance.

Nate raised his eyebrows. "Trevor, come here. Scotty, show him your calculation." Scotty had calculated how much DC electricity could be produced by a tub of seawater the size of our fire pit, using harmonic oscillation to generate Higgs Bosons that would transfer energy to fractal capacitor circuits which ascend in capacity: enough electricity to run the city of Sydney, Australia for over a thousand years.

"How do you boys know this?" Professor Hansen asked them in a serious tone.

Trevor glanced at his three physics buddies and Scotty. They all nodded, signaling he could keep going. "We were camping in the bush west of the Blue Mountains last year, way back in the bush! On the third day we met some tall blonde people in white robes who explained this all to us. They were beautiful and smart—they said they lived underground in Tibet."

"Tell us more," the Professor requested, rubbing his chin.

"There were four of them, two men and two women, about six-and-a-half feet tall with dark blue eyes. They wore silver suits under their robes."

"Nordic blonde ETs," I mentioned to Graham.

Graham looked at me. "This has me tied in a knot." When hard-to-comprehend scientific information junctions with the unknown and sounds impossible, Graham gets an upset stomach.

"Settle down, Graham. These boys likely met the good ETs," I assured him, hoping I was as sure as I sounded.

Everyone seemed to feel the need to pause. Trevor and his three partners picked up the Frisbee and began throwing it to each other, the dog joined in. When Nate got involved, he offered the dog another cookie. That got Helen to smile. . ..

Postscript

Tall Blonde entities, also called Nordics, look very human—some say super-human. They belong to one of five groups of ETs that are part of the alien presence on earth. Numerous documented contacts with humans have occurred. These ETs are quite handsome and benevolent according to many accounts. Purportedly, they are here to help protect us from ET groups that are unfriendly—specifically a reptilian race of underground ETs that claim they've been on earth for millions of years.

We didn't pursue any deeper discussion that night on the beach. It didn't seem appropriate, and I did have an early flight to Auckland in the morning. I met with Professor Hansen in Seattle five months later at a chemical-society conference where we further discussed the situation. The professor had begun working with Trevor, Nate, Scotty, and Helen in an effort to gently perturbate the Higgs Field and capture the energy with frequency coherence by using modified Van de Graaff generators and Marconi coils. The resulting high-voltage DC current charged a series of supercapacitors where it was stored for use. Producing free electrical energy was their mission, understanding how to dig it out of the Higgs Field with help from God Particles was the key. Trevor's three accomplices helped out with "the project" on occasions but had their own PhD assignments and research to complete. Joni got married and had her first child and took a break from physics. Nate adopted the dog and taught Helen how to surf.

The 2013 Nobel Prize in Physics was awarded to Peter Higgs and Francois Englert for the discovery of the Higgs Boson. The God Particle was proven real at a confidence level of five-sigma.

Oahu, Hawaii — Honolulu with Diamond Head and hotels

Chapter Ten

The Deep State Divide
Paradise Almost Lost

Prologue

The unelected shadow government in America has been called by various names: The Cabal, the Illuminati, the Military Industrial Complex, the Global Elite, and the *Deep State,* which is how I'll refer to it in this story. This shadow government is not a single entity but a clandestine consortium of covert networks populated with powerful individuals from government, critical industries, and finance.

Without consent of the governed, this Deep State forces the world to follow their plans. To them it's a big game of Monopoly® that includes the games of Clue® and Risk®. Exerting control over major world affairs is their mission, and clandestine activities are used whenever required: wars, terrorism, recessions, depressions, and assassinations top the list of techniques employed. Although the Deep State's malevolent majority generally rules, benevolence does exist within the minority. This, as you might imagine, can create viperous disagreements that largely go unnoticed "above ground." The complexity of the Deep State crosses all borders and, somehow functions worldwide beyond definition in any conventional sense. But it surely exists. I can say this with confidence, because as a scientist that has travelled the world searching for natural solutions to formidable problems in agriculture, human health, and the environment, I have met many brilliant individuals and encountered unexpected challenges that, together, tutored me through myriad realms.

The "modern" Deep State originated during World War II, when a breakaway group of Nazis escaped to South America and Antarctica, and reinvented themselves as The Fourth Reich. A significant contingency of Aryan Germans with strong military and industrial backgrounds—intoxicated with Nazi brainwashing—accelerated their move out of Germany in 1944 when it became obvious the war in Europe would be lost. This Deep State proceeded to evolve over the next seven-plus decades following the U.S. victory in Europe on V.E. Day, May 7, 1945. Its complexity grew as the malevolent and benevolent forces competed for control. After the Kennedy assassination in November 1963, Deep State entanglement within the U.S. military, defense industry, and political process became inextricable and the hawkish factions caused the death of fifty-eight thousand U.S. soldiers in Viet Nam.

The presidential election of 2016 was a game changer. Today, in 2019, malevolent forces tend to still dominate the Deep State and perpetual wars continue; however, there appears to be an increase in benevolent luminosity at the end of the tunnel. Good photons! For the sake of humanity, this light needs to intensify.

Hawaiian Islands, January 2018

On January 13, 2018 at 8:07 a.m., alarm messages on cell phones throughout the Hawaiian Islands signaled that a missile attack was imminent. From the Big Island to Kauai panic spread as people read the text message: "BALLISTIC MISSILE THREAT INBOUND TO HAWAII—SEEK IMMEDIATE SHELTER. THIS IS NOT A DRILL." The U.S. Pacific Command (PACOM) had issued an incoming missile alert for Hawaii.

CBS and NBC television affiliates interrupted regular broadcasting with a more detailed message. Twitter simultaneously reported: "Hawaii Civil Authority has issued A CIVIL DANGER WARNING for the following counties or areas: Hawaii at 8:07 AM on JANUARY 13, 2018, effective until 6:07 PM."

The text on a KGMB TV video clip read: "The U.S. Pacific Command has detected a missile threat to Hawaii. A missile may impact on land or sea within minutes. THIS IS NOT A DRILL. If you are indoors, stay indoors. If you are outdoors, seek immediate shelter in a building. Remain indoors well away from windows. If you are driving pull safely to the side of the road and seek shelter in a building or lie on the floor. We will announce when the threat has ended. Take immediate action measures. THIS IS NOT A DRILL."

Emergency sirens rang out at Pearl Harbor's Naval Station; personnel began evacuating. There were no indications this was just a drill. Many who attempted to flee Pearl Harbor and Honolulu—anticipated targets for the attack—got stalled in morning traffic.

The unimaginable, unthinkable destruction of paradise was about to happen again after eighty-one years. Thirty-eight minutes of sheer terror passed before the emergency alert system cancelled the warning—it had been a false alarm.

The mistakenly-issued Ballistic Missile Alert (BMA) was blamed on a lack of management controls, obsolete computer software, and human error. It was claimed there was no missile—a planned BMA drill simply went awry. The Hawaii Emergency Management Agency (HI-EMA) concluded that "someone pressed the wrong button."

But then eyewitness sightings of an aerial explosion started coming in, along with other contradictions to HI-EMA's report, supporting indications that there *had indeed* been a missile—an actual attack in progress.

It was no mistake, PACOM had issued an authentic alert and had simultaneously contacted the White House for approval to take retaliatory action against foreign vessels in the area. The White House replied, "Negative" and demanded more information.

A group of tourists in a boat off the coast of Maui reported seeing something get blown out of the sky around 8:00 a.m. on January 13. "It looked like a meteor and lit up the entire sky causing a big boom," one tourist exclaimed. Others made similar comments.

Multiple reports to the media and civil authorities in Kauai described a bright aerial flash and thundering explosion around 8:00 a.m. on January 13. One report was aired on a local TV news channel one time before being pulled; ditto on Oahu. There was a plethora of such reports in the following days, with people claiming that men dressed in dark suits and sunglasses had threatened them if they talked to the press. Sightings of an aerial explosion suggested a planned attack that was somehow thwarted.

Fact: The U.S. Military is unable to shoot down intercontinental ballistic missiles (ICBM) or submarine launched ballistic missiles (SLBM), particularity during their descent from suborbital space. Neither the antimissile range on Kauai nor the Navy's Aegis anti-missile defense system on ships have the capability to shoot down an incoming hypersonic ICBM or SLBM. So, who—or what—initiated a nuclear attack on Hawaii: a foreign state, or was it a false-flag attack? Classified investigations were initiated by the U.S. Air Force (USAF) and U.S. Navy (USN).

Guam, one year later

Navy Commander Trevor Smith was sitting alone staring at his laptop in the noisy pub, when Air Force Colonel George Simon, came in the bar's back door. It was a warm, January evening in Micronesia on the U.S. island of Guam. The bar was eight miles from Andersen Air Force Base, the largest U.S. airbase in the Pacific; the two officers were involved in a highly-classified joint investigation.

"You found a crowded bar, Commander—perfect place." The colonel grinned. The late-forties, blue-eyed, sandy-haired ex-submarine commander clicked the save tab on his Word® document then bumped fists with the colonel. They didn't salute when alone.

"The Stones are playing my song," Trevor remarked. *I Can't Get No Satisfaction* blasted out of the bar's stereo system, vibrating his Gibson martini.

"We'll have to fix that soon," the colonel asserted.

"Sooner's better than later! Did you have a smooth flight from Pearl?"

"Smooth as silk on the C-32; much better than the old turbo props." The colonel ordered drinks and got comfortable in their sequestered location under a large blue-marlin mount on the wall behind them.

After slugging down a shot of Crown Royal followed by a long drink of cold Lowenbrau, the colonel opened a worn leather case and took out a blue folder marked "Top Secret." Trevor turned his laptop off and closed the lid.

"Did you get the Christmas card I sent?" George asked Trevor.

"Yes Colonel. Thanks for the informative message." Around the holidays, greeting cards were a secure way to send messages. Billions are sent every November and December in the USA—too many for the Deep State to monitor.

George nodded and opened the folder. "The missile was not a test or an accident!" He declared.

"Nuclear payload?" Trevor surmised.

"Confirmed. Two W-76 warheads—each with a one-hundred-kiloton yield."

"Both aimed at Pearl?"

"One at the harbor; one at downtown Honolulu. The Deep State wanted a major kill."

"So North Korea, Russia, and China have been completely ruled out?"

"Completely!" George asserted. "Hawks in the Pentagon were furious. They wanted to vaporize North Korea regardless of what we found."

"Holy Shit!" Trevor spilled his martini." That could have been one of our Trident D5 missiles that went missing just over a year ago."

"It was launched from a Dark-Fleet submarine," the colonel acknowledged.

"A CIA asset!" the commander asserted. The motive was clear now: Nuke Hawaii, blame North Korea, end any peace efforts, and start World War III—then advance the movement to take control with a one world fascist-socialist government. "The Deep State is split; the CIA's fascist-socialist faction operates the Dark-Fleet."

"It's amazing those ancient Nazi subs still function," the colonel responded.

"They've been upgraded but, that notwithstanding, you're right. It's still amazing," The Commander tried to wipe up most of the martini with a cocktail napkin.

"It's a good thing the war ended when it did! The Nazis had us by the short hairs, technically." The colonel watched the couples on the crowded dance floor. He pinched his chin while holding his elbow and leaned back. It had been two years this month since his wife had passed. He shook his head and looked at Trevor.

"Sorry for the diversion," He moved the blue folder away from the spilled martini. "Are those cocktail onions sour?"

"Not too. Try one." George picked one off the table next to the spilled drink and ate it.

"They feed the good bacteria in your gut," Trevor remarked. "I always request two in a martini." Then he picked up the second onion and ate it. Serious times require trivial diversions; Trevor got up and shouldered his way to the bar to get another round of drinks. A female lieutenant he had once danced with gave him the eye.

"Here's the cure for your queasy gut." Trevor handed George a shot of Crown Royal with two onions in it. "The missile had to be launched close enough to North Korea to match a likely trajectory—D-5s have a range of seventy-five hundred miles when carrying two W-76 warheads; Korea is forty-six hundred miles from Honolulu."

"Affirmative," the colonel acknowledged. "It was launched northeast of Japan over the Kuril Trench."

"Like we suspected six months ago!"

"Only, unequivocally substantiated now." George patted his USAF identification pin. "Our Special Ops group had to go beyond the approved book of investigative tricks to get the facts."

Trevor smiled; he didn't pat his USN command insignia. "Wish I could say the same for the Navy's Intel—they haven't been much help. Certain facts always seem to escape discovery. How the hell did the USAF Special Ops ever shoot that missile down? It was in descent travelling over eleven thousand miles per hour."

"Are you ready for this?" The colonel glanced around to make sure only the marlin hanging above them was within hearing distance.

"Ready!" Trevor took a drink of his second martini.

"The missile was destroyed by an advanced laser cannon aboard a highly- classified, rectangular military platform in geosynchronous orbit five-hundred miles above Earth. The laser beam smacked the hypersonic SLBM on descent. Laser beams travel at the speed of light."

"It had to be instantaneous and powerful to prevent a nuclear detonation." As a nuclear-submarine commander, Trevor understood this—it was one reason antiballistic-missile defenses risked serious collateral effects.

"The laser canon's million-joule microwave pulses destroyed the missile and its warheads within microseconds of striking the SLBM—not enough time for a thermonuclear nuclear chain reaction to develop. Any residual radioactive lithium deuteride is on the bottom of the Pacific Ocean," George confirmed [1].

"Astonishing!" Trevor remarked. "Did our Nordic friends help track the missile?"

"We tracked it using the Maui Space Surveillance Site (MSSS) on top of Haleakala. The Nordics checked our computers before we deployed the laser. Photos of the CNN broadcast during and after the warning revealed multiple UFOs in the sky. Too much of a coincidence to claim no involvement by them—they had our backs." The two officers continued discussing additional details of the missile's destruction. . ..

"Congratulations, George!" Trevor was very impressed and thankful; he had family in Hawaii. "I assume the Special Ops men that were involved retired early after signing an updated non-disclosure agreement?"

"Yup, that's the way the ball has to bounce. I believe you know what's next for us."

Trevor looked around the bar that was near Guam's Tumon Bay. It was still lively, noisy, and well attended by the American military, featuring an abundance of naval officers. A U.S. nuclear aircraft carrier had docked in Guam the day before, which always meant good business. Convinced nobody could hear them, the commander answered the colonel:

"We need to investigate the Deep State and expose the malicious dark side that was involved in launching the false-flag attack on Hawaii. Make arrests and incarcerate the bastards for eternity."

"They're the same excrement-laden vermin that ordered the 9-11 attack in New York. They beat us with that tragedy, but we saved Hawaii. The world is at the stage where we may need your *Blue Bird* friends to come forth and provide additional help." The colonel's face was red with concern. "Are they still willing?"

"Still willing, Colonel. They're ready to help our helpers."

"So the *Blue Birds* would help the Nordics help us?"

---

1 Lithium deuteride can replace plutonium as the "spark plug" for thermonuclear hydrogen bombs.

"Correct. We're not advanced enough to work directly with them," Trevor smiled, "we're down a bit too far on the evolutionary scale."

"I hope they understand we have the potential to advance," the colonel declared.

"They do! I assure you. They've been around, hiding in the space between the spaces—trying to sort which of us humans are the good guys."

So, with Led Zeppelin blasting *Stairway to Heaven* on the stereo system, the two military officers paused to stare at the marlin hanging above them. They appeared to share the same thought: *How nice it would be to fish for giant marlin in Micronesia's tropical seas, instead of having to confront a dark force that shouldn't exist.*

At this stage, both the colonel and commander knew of the split in the Deep State and its two opposing forces. In the mid-1950s there was only one force containing mixed remnants from the war, including Antarctic Germans, U.S. military leaders, advanced weapon scientists, CIA counter-intelligence specialists, and defense contractor CEOs. President Eisenhower referred to the group as the M.I.C.—the Military Industrial Complex. He had been pressured into accepting it, based on the significant technological backwardness of the U.S. compared to the Antarctic Germans and their alien supporters. At the end of his second term, Eisenhower issued a warning in his last televised speech on January 17, 1961. The following is a quote from that speech:

"In the councils of government, we must guard against the acquisition of unwarranted influence, whether sought or unsought, by the military-industrial complex. The potential for the disastrous rise of misplaced power exists and will persist. We must never let the weight of this combination endanger our liberties and democratic processes."

By the mid-1980s the evolutionary process had created a new diversity of individuals comprising the Deep State, including, in all seriousness, other-worldly participants (ETs) or their proxies who were unquestionably at odds with each other! This formed the basis for the split: A Bright Force diametrically opposed to a Dark Force.

The Bright Force included USAF and USN leadership, supportive members of the defense industry, positive elements of the intelligence community (mostly DIA), prominent scientists, and the *Nordic ETs*. Whereas the Dark Force was comprised of various fascist-socialist factions: Second generation Nazis of the Fourth Reich, representatives from the CIA's clandestine service, defense industry globalists, and proxies for the *Reptilian ETs*. The Bright Force supported U.S. nationalism, the constitution, bill of rights, and individual economic freedoms. The Dark Force promoted a one-world fascist-socialist government with plans to reduce the USA's power, status, and population, and eliminate individual freedoms. The relationship between the two sides, to put it mildly, had become one of vicious hostility.

It's important to understand that the Deep State has no formal charter, operation manuals, or provable acknowledgements of its existence that I am aware of. It is difficult for most people to believe that such an "organization" could actually exist and that its activity could be hidden, largely in plain sight.

Berlin, Germany, January 1996

When I arrived, there was snow on the *Kurfustendamm,* or *Ku'damm* as local Germans call Berlin's renowned boulevard. The Berlin Wall was gone thanks to Ronald Reagan, along with the sign outside the Brandenburg Gate that warned: *ACHTUNG: Sie verlassen jetzt* WEST-BERLIN (ATTENTION: You are now leaving West-Berlin). I remembered it well from my first trip to Berlin in 1968 as a college student. The way into East Berlin for U.S. citizens was through Checkpoint Charlie—and going in came with warnings from American authorities at the checkpoint: "People will beg you to sell them your shoes and coat. Don't! You will be arrested and put in detention indefinitely by the East German police. And we won't be able to help you!"

I entered East Berlin reluctantly back then, with two college friends. We quickly comprehended how bad things were under socialism. Quality shoes and clothing didn't exist for the common man—only for the socialist party elites. A woman who had been a high school teacher stopped us and explained, "We are all poor, cold, and hungry with holes in our shoes, bugs in our clothes, and only acid in our empty bellies. It's the great equality that socialism promises which is actually a hellish, unproductive sameness! But, if we reject it, and run to climb the wall into West Berlin to find freedom, we will be shot dead by the East German police. T*he concept of equal misery for all must be enforced with walls that keep people in and guns that kill those trying to get out.*" I'll never forget her words.

I gave the woman my insulated leather gloves which she immediately hid under her thin, gray coat. I told my two college friends: "We must never allow this to happen in the United States of America." It was my first real-life exposure to socialism. The people were under total government control—they had no free speech, no private property, no self-defense, no right to assembly . . . only government approved books and magazines to read, unreasonable travel restrictions, and rationed food and medical services.

That was an important lesson for me in 1968. Major improvements occurred after Berlin's reunification in October of 1990. Life-strangling socialism ended, and economic and individual freedoms returned to the citizens.

In 1996 I visited Berlin to follow up on an offer I couldn't refuse. My good friend, Karl Schultz, had invited me to interview his ninety-year-old grandfather regarding his connection to events that occurred in the Wehrmacht (German military) during World War II. Karl was my age and had a degree in mechanical engineering. He knew that one avocation of mine was studying the hidden history of World War II.

His grandfather, Herr Hans Schmidt, was a retired chemist who was finally ready to share his secrets with an American chemist. It had been his experience that chemists respected each other, and it was a dream of his to meet one from America. Karl told him he knew one—me—and would be happy to translate during an interview. I took the train from Copenhagen, where I'd had business meetings the previous week, to Berlin to meet with Karl and his grandfather.

The old man's story started in the 1930s in Nazi Germany, before Hitler attacked Poland in September of 1939, starting World War II. Three powerful industrialists, I.G. Farben, Gustav Krupp, and Fritz Thyssen, who had funded the evolution of the Third Reich, wanted the advanced Nazi war-technology that was under development—particularly laser cannons, antigravity aircraft, and gravity warping propulsion systems—moved out of Germany to a more remote and safer location. Queen Maud Land in eastern Antarctica had been explored in the late 1930s and Nazi flags planted. Hitler agreed to begin moving his beloved flying saucers and experimental ray guns to the infinitude of ice and frozen tundra of Antarctica—but it couldn't be done in a rush; he had a war to fight.

During the extreme winter of 1941-1942, Hitler's military blunders in Russia froze the unprepared Wehrmacht, forcing a retreat, putting the Third Reich in danger of losing the war. It was only a matter of time in the minds of the powerful industrialists, so they insisted that Hitler pick up the pace of moving the Reich's advanced weapon systems. A plan was developed, and the Nazis expanded their fleet of large transport submarines necessary to covertly relocate the weapons and key personnel. It would be a formidable 10,500 mile underwater boat ride from Germany to *Neuschwabenland*, Antarctica—multiple voyages would be required.

"Herr Schmidt, what was your involvement with the advanced weapons?" was my first question while sitting in a comfortable armchair in his grandson's living room on the fifth floor of an apartment building that had miraculously survived the war one block off the *Ku'damm*. Herr Schmidt, who was lying on the couch, understood the question and waved off the need for Karl to translate.

"I was one of three chemists responsible for propellant formulations, for bombs that could fly."

"You mean rockets,"

"*Ja, Raketen.*" Hans laughed; he liked the English word better.

"Big ones larger than the V-2?"

"*Ja, Der Führer* wanted bigger *raketen* to blow up New York and Washington."

"Were you involved with other war technologies?" I shifted positions in the comfortable chair to see Hans' facial expressions better.

"*Ja,* many things were happening in two places, Peenemunde and Nordhausen with not much secrecy. The war made us scientists crazy, even when protected underground." The old man rubbed the left side of his head, next to a bad scar where his ear was missing.

"He lost his ear in the war," Karl commented.

"It was the best thing that happened to me in the war!" Hans declared, sitting up. "Whenever a Nazi SS officer ordered me to do something, I played deaf. The SS had orders from General Hans Kammler not to bother us scientists. The Fuhrer personally put Kammler in charge of all advanced war technology. The general didn't want the SS bothering us. When they tried anyway, I just shook my head and pointed to my missing ear."

"Did Werner von Braun report to General Kammler?"

"*Jawohl* (yes)."

"Which weapons did they move first?" This interview was already turning out to be more than I had expected.

"All prototype antigravity discs were first; some had been moved earlier. We were very careful to completely cover the flying bells and discs with protective tarps. The big ones had to be disassembled to fit into the submarines. There was a lot of mercury to move."

"Tell me more about this." I unconsciously started rubbing my left ear.

"Antigravity was the biggest secret. There were two complementary mechanisms: One involved spinning hollow metal rings containing mercury in a plasma state at very high speeds: fifty to seventy thousand rpm. They were under great internal pressure and required cooling to keep from overheating. We called it *fast-mercury;* it negated earth's gravity with its own gravitational force and reduced the weight of any craft by ninety percent. The other mechanism used huge capacitors charged with millions of volts to create electrogravitic propulsion: gravity warping or super-thrust."

"Did you work with antimatter propulsion?"

"Too dangerous! It works when you know how to contain it—a large laboratory in Bavaria blew up trying to contain it in 1942. Engineers were shooting protons at an isotope of element 115 in a vacuum. When done correctly, this produced element 116 and released antimatter. It had to be done in a specially designed, tubular reactor that was very dangerous to operate . . . I knew one of the chemists who was killed in the explosion. He told me, the last time I saw him, that element 115 was made in outer space when two stars rotated around each other. To be stable it needed many neutrons in its nucleus, requiring tremendous energy to form."

Wow, I thought to myself; these were all new details for me. "Were the Nazi discs—the flying saucers—capable of going into space?"

"Not in 1945. Not from any made in Nordhausen or Peenemunde, don't know if any of those made down south could—maybe the Vril machines could. Ours could fly hypersonic for five hours at Mach five; then required recharging. We needed more advice on how to improve the technology."

"What in particular was missing?"

"A way to tap into vacuum energy—its fluctuations. And trap and store it; the capacitors on the super-thrust drives required frequent recharging."

"Couldn't the ETs provide that information?" I shifted my position in the chair again. I knew that vacuum energy referred to the universal energy field that penetrates all space. Some call it Zero Point Energy. Fluctuations are recognized by physicists but it's understood that trapping and storing such energy it is not possible. *Just like everything else we were discussing—not possible.* The thought made me smile to myself.

"General Kammler told us we would be working with different aliens in Antarctica—lizard men. And they would tell us!"

"Which ETs did you work with in Germany?"

"I only saw them once: Tall grey characters in robes. They glided across the floor, their feet didn't touch. There were also some shorter ones with six fingers and toes that the other scientists mentioned, but I never saw them. I was told they were flight instructors for the discs."

"These are now called the Tall Greys and the EBENS; the lizard men are the Draco or reptilians." Hans nodded; he understood what I'd said. It was obvious the Tall Greys and EBENS only taught the Germans enough to get them started with antigravity discs in the earth's atmosphere.

"There may have been a third group of tiny guys—not much over a meter tall—with pear shaped heads and large slanted eyes; they worked underground in Nordhausen and were in a photograph I saw," Hans added.

"Those are organic droids or Little Greys; they're the worker bees for the Tall Greys." I grinned.

"Who are the Nordics, then?" Karl asked me.

"They're reported to be a robust, healthy ET race, six to seven feet tall with blonde hair and blue eyes; their skin color is typically tan. Both male and females exist. They look much like us humans and are quite advanced. Sometimes referred to as the Tall Blondes, they purportedly originate from the Aldebaran Star System."

Karl looked puzzled. "Please understand, Karl, I've never met a living ET that I know of, just like I never met George Washington. It comes down to knowing what evidence and who to believe."

The three of us talked more about the different ETs and how conflicting much of the information is; knowing how to sort out the authentic from the bogus is a formidable challenge.

Hans described the command structure under General Kammler and said he wasn't invited to Antarctica because of his missing ear.

We took a break for lunch. Karl drove us to a sausage factory with a restaurant attached—near the Spree River. They had a variety of finger-sized, spicy pork sausages on strings. Those fortified with garlic and barley were great. Han's didn't eat many but managed to finish 500 milliliters of Hofbrau, then took a snooze sitting straight up in a wooden chair. Beer hadn't changed in *Deutschland*; it was still defined as a food.

"I believe everything my grandfather is telling you, but how can it be proved?" Karl asked me.

"Let me try to explain while your grandfather takes a nap." The restaurant wasn't busy so we stayed at our table in back—looking out the window at the Spree River had a calming effect. Good thing, because telling the following story tended to elevate my blood pressure.

"The United States government signed an agreement to coexist with the Antarctic Nazis in 1955. President Eisenhower was directly involved. It was complicated and involved meetings with the Nazis, and separately with at least two different ET groups. When Ike understood how far behind the U.S. was technologically, he had few options. Remember, Germany surrendered on May 7, 1945. Hold that thought.

"In August 1946, fifteen months after Germany surrendered and the war ended, President Harry Truman sent a task force of thirteen warships, including the flagship Mt. Olympus and the aircraft carrier USS Philippine Sea, forty-eight hundred men, one hundred fifty aircraft including P-51 fighters, six helicopters, six flying boats, and two seaplane tenders to Antarctica to investigate and clean up any last remains of Hitler's Third Reich if there were any. Called 'Operation High Jump,' it was billed as a mission to search for coal deposits and update mapping. To say such a mission was highly unusual and irregular would be a gross understatement.

It operated as 'Task Force Sixty-Eight' and was commanded by Rear Admiral Richard H. Cruzen, USN, while Rear Admiral Richard Byrd supervised the overall mission. All this is well documented.

"British Naval Intelligence had reported a fleet of Nazi submarines and strange circular aircraft in eastern Antarctica in the summer of 1946—at a place in Queen Maud Land that the Nazis called *Neuschwabenland*. This was not known outside the White House, and a small number of Navy admirals and Intelligence officers. Truman (then president of the U.S.) put a lid on it—the buck stopped with him.

"To make a long story short, Task Force Sixty-Eight got its ass kicked by the 'Fourth Reich.' There were American fatalities and damages to the ships—not at all trivial, exact numbers are still classified. The Nazis attacked our fleet using flying saucers armed with laser guns. There are death bed testimonies supporting this! The U.S. P-51s had no chance. The hypersonic saucers came out of the sea and grossly out-maneuvered and shot down the P- 51s. Pilots were killed. Ships were damaged and one destroyer was sunk along with two other boats; quite a price to pay for postwar coal exploration and a mapping expedition.

"The Germans could have totally destroyed our fleet, but instead sent us home six months early, limping badly. The Nazis were at least two, possibly three decades ahead of us technologically. Eisenhower had to agree to their terms for coexistence.

"The U.S. Navy covered it up with a multiplicity of excuses. Everybody was still celebrating the end of World War Two in Europe—V. E. Day, May 7, 1945. The 1946 fatalities in Antarctica were reported as accidents. No coal was discovered. No fights with Nazi flying saucers were mentioned. Much of the paperwork and logs were burned and magnetic computer tapes 'accidentally' exposed to magnets rated at fourteen thousand Gauss. Cleanup was easy in those days.

"Admiral Byrd died unexpectedly, sometime after the return of Task Force Sixty-Eight; he had given one too many speeches declaring he witnessed circular aircraft that could fly from the South Pole to the North Pole at great speeds and that the USA homeland was in danger. The U.S. first introduced a hypersonic jet, a jet that can exceed Mach 5, in the 1960s—the X-15." While I explained this to Karl, Herr Hans participated in a high-decibel snoring contest with another old man, a total stranger who had moved his chair behind Hans. Karl and I didn't realize this until we heard their unusual harmony.

"Also Karl, consider this: Fifteen-hundred of the best German rocket scientists were brought to the USA after the German surrender in May 1945, along with Dr. Werner von Braun, their leader. It was called Operation Paperclip. They were given residence visas along with comfortable accommodations and allowed to continue their work in Florida and New Mexico. Although mostly rocket specialists, many were well aware of what their compatriots were up to in Antarctica. Information was shared. Fifty years later their testimonies are still classified Top Secret. Go figure."

"So, the Third Reich was still in business in 1946 and well beyond?" Karl stated.

"Absolutely, except they now called themselves the Fourth Reich."

"And the reptilian aliens, the lizard men, disclosed critical technical secrets once the Nazis were comfortable in the earth's deep freeze?"

"From what your grandfather has told us, yes, that appears to be the case. And it correlates well with what I know from other sources. The reptilian aliens came from the Draco star system originally, and claim ownership of Earth. Supposedly they have been underground since the dinosaurs—over sixty-five million years ago—whereas Lucy, humanity's most distant, upright-walking ancestor, evolved two million years ago. Big difference!"

Hans woke up and scratched his ear. "I'm ready for more questions!" he declared in English. Suddenly he looked twenty years younger; the nap must have had an anti-aging effect.

"*Willkommen zurück!*" I said to him in German. He grinned, showing most of his nonagenarian teeth. "How much did Hitler know about all this?"

"He knew, of course—how much is debated. He knew more about what his generals were doing than his admirals. The Supreme Commander of the U-Boat fleet, Karl von Donitz, probably knew the most. Germany had over one thousand U-Boats—they knew every undersea hideout on earth."

Hans paused for a moment then added, "Hitler knew not to dispute the Farben, Thyssen, and Krupp dynasties; he was a puppet for the German industrialists. For the Nazis to rule the world they would require exclusive access to all their advanced-weapon systems. The importance of this went beyond Hitler. It meant moving out of Europe and getting friendly with the right aliens. Hitler had to follow orders."

"So, Hans, you should know what I'm about to ask." I grinned.

"Did Hitler escape to Antarctica?" Hans replied, looking at me. Karl listened intently.

"That's it!" I said. "That's the sixty-four-million-Euro question."

"I really don't know. There's evidence that suggests he did. He had several doubles; one was a very good copy of him." Hans scratched around his missing ear.

"Can you add any more detail about the lizard men's activities in Antarctica?" I asked.

"I was told by my chemist friends that the lizard aliens taught us how to recharge our flying discs continuously using vacuum-fluctuation-energy. All I know is it required copper coils wound in unusual ways that revolved around magnets which were connected to large capacitors.

"Also, heat was generated by a mechanism that involved running electricity from the capacitors through a metallic device: a bunch of tubes that heated up. When this heat was converted back into electricity with a thermal-electric converter, we got more electricity than we started with. My chemist friends were more impressed with this than anything else.

"They also redesigned our laser guns. Ours were disturbing some critical geometry in the craft that interfered with the antigravity mechanism. And, very important, they showed us how to produce unique metal alloys—ones that could help position and rotate the craft. Magnesium and Bismuth were important elements to combine in layers that actually became part of the craft's electrical system, not just its outer fuselage. The lizard men had underground factories below Neuschwabenland where this all could be done. One of the bomb chemists told me it was pretty nice under the ice—twenty degrees Celsius."

"Did the Reptilians—the Draco—threaten Hitler?"

"I think, yes. I was told the Draco were the only living creatures he feared. Hitler tried to slow things down with them. He was afraid they would take over. And if it didn't go his way, he would leave Neuschwabenland for Columbia or Argentina until things settled down. Apparently, the Reptilians had been hiding under the ice in Antarctica for eons; they didn't do well above ground in sunlight." Hans rubbed his scar again. "From what you explained about the battle at sea in 1946, the saucer technology problems had been solved by then."

Could Hans have understood what I explained to Karl while he was sleeping and snoring?

"So, the Fourth Reich was born in Antarctica?" Karl summarized. "They're literally the Deep-Freeze State." I laughed.

"There has been significant evolution in the Deep State in fifty years, particularly in the past twenty," I assured him.

Karl had ordered three fresh steins of Hofbräu, each sporting two full inches of foam. A muscular lady, one you would never pick a fight with, carried all three of the golden München brews over to us with one hand.

"Good timing!" I nodded to Karl, while smiling at Brunhilde.

"*Prost*!" Hans raised his stein. Karl and I did the same. It was a welcome pause.

I looked at Karl. "A parallel, unelected government with plenty of scary entities, both human and non-human, has been brewing for a long time. I would give back one of my larger marlin catches to have been a mouse at their early meetings. There had to have been a nasty sorting-out process. The result was a split Deep State: A Bright Force versus a Dark Force—freedom versus enslavement.

"What's stopping the Reptilians from taking over?" Karl asked.

"Seven billion humans, for one thing. It's estimated there are fewer than ten thousand reptilian ETs on earth," I answered.

"And humans have nukes! The Galactic Federation won't permit nuclear weapons in space, period. They disrupt the frequency-coordinates for locating star gates and portals. Nukes punch holes in spacetime, so to speak.

"Also, the Reptilians hate the Nordics, they understand that the Nordics will side with the Bright Force in any war."

"So the lizard men have taken control of the fascist-socialist faction of the Deep State—the Dark Force—in an attempt to facilitate their takeover of the world!" Karl summarized. "It's them against the freedom faction—the Bright Force, the Good Guys—us."

"Correct. That's the game that dominates mankind," I responded. The three of us paused to finish our beers.

Hans, wide awake, had been thinking during the pause. "It's necessary for people to understand that the Reptilians, the lizard men, must be tightly connected with the far-left socialists. Once the world's population has been reduced by socialist-prompted civil wars, the ten thousand Reptilians will require help controlling the surviving humans." Karl seemed surprised by his grandfather's astute remark.

"Are the Nordics superior to the Reptilians technically?" Hans asked me.

"From what I understand they're about equal—these two ET races have a history of fighting each other to a standstill."

"*Beeindruckend!*" Karl proclaimed. "These are certainly serious times!"

Hans asked Karl to order him another 500 ml of Hofbräu. "That will surely put you back to sleep, *Opa*."

"No it won't!" Hans said stubbornly. Karl quickly ordered another stein for him.

"So, where do we go from here?" Karl queried.

"Let's keep talking and see if we can figure that out . . .." Herr Hans insisted.

The three of us stood up and walked over to the large window that overlooked the Spree River. It was frozen along both shorelines but had open water out in the middle, gray water on a gray day in Berlin. Two white swans fluffed themselves in the windy water; they were cold and trying their best to paddle across to the other side. Snow began falling—there was purity in the whiteness . . ..

Postscript—2019

The description of the attempted nuclear attack on Hawaii in January, 2018 is factual—it happened. Thank God it was thwarted. The U.S. government has not confirmed the missile, but the U.S. Coast Guard spent the rest of the month of January 2018 searching the Pacific Ocean from Maui to Kawai. Not a routine endeavor.

The meeting between a USN naval commander and a USAF colonel in Guam a year later was explained to me by two separate, trustworthy sources; both had served as officers in the U.S. Military.

Hans Schmidt was ninety when I interviewed him in Berlin in 1996—long before the near-tragic event in Hawaii. I believe he believed everything he told me. Hans passed in 2001. Karl moved to Australia in 2004.

The fact that a Nazi breakaway group established itself as the Fourth Reich under the ice in Antarctica in the 1940s is well documented in German. That an association with Reptilian aliens made this possible can be questioned; I expect further evidence to come out confirming this soon. The Fourth Reich is still represented within in the Deep State.

The U.S. has been operating at least two Secret Space Programs (SSPs) since the 1980s: The Navy's Solar Warden program and another managed by the USAF Special Operations. In fact, these two programs were secretly competitive with each other until recently and will soon be combined under Trump's new Space Force.

We have sent astronauts to the back side of the moon, to Mars, Ceres, Europa, and beyond. Many of the UFOs sighted are our craft, but not all of them. There are at least five, likely seven, ET races currently on planet earth. During the course of my business and science travels to over fifty countries, I have come to realize this is an accepted reality among many people. All the Presidents since FDR were briefed on it to varying degrees. President Kennedy's plan to disclose the alien presence is what got him killed—information gained from the Freedom of Information Act leads directly to this conclusion. You can find it on the Internet.

Today in 2019, advanced versions of our triangular, antigravity, TR-3B spaceships frequent the skies and come in three sizes: small, medium and large. The large ones—seven-hundred-plus meters long—even give UFO buffs pause. I saw one in Brazil a decade ago.

I have no information on how many galactic-capable spaceships the USA has today, either on earth or elsewhere in the solar system or greater cosmos. My guess is many! We are now a cosmic force among the aliens and many of them are afraid of us.

There are wormhole shortcuts in space-time that can take our spaceships to other solar systems in the galaxy in a matter of hours, or days. Certain ET groups have helped us reinvent our understanding of physics and non-linear, multidimensional mathematics—an important step. The subsequent production of antigravity spacecraft utilizing non-rocket propulsion involved successful integration of the following: (1) Reverse-engineering of various ET craft—of which there are many; (2) Direct communication with ET entities for explanation of critical details; (3) Multibillion-dollar pilot facilities for testing components and prototypes. The need for oversized blackboards filled with complex mathematical equations was less important. Working with test models was strategic!

The biggest mystery for me—confusion with bogus events notwithstanding—is how this all has been kept secret for so long. NASA, which is essentially a front organization, has provided cover by using low-tech rocket propulsion projects between the moon and earth since 1969. Since then, hundreds, likely thousands, of earth men and women have taken and are still taking part in the USA's Secret Space Programs (SSPs). Disclosure of this is on the horizon; a plethora of hard evidence is ready for show and tell, too much to keep hidden.

The Dark Force wants this disclosure to happen slowly—over decades—permitting a slow evolution into Free Energy and space travel while maintaining tight control over the population. The Bright Force wants it out ASAP, with individual freedoms preserved and worries about climate change, pollution, food production, and much more solved by limitless Free Energy.

Looking at the two major ET races involved in this struggle, Nordics versus Reptilians, it's very hard to predict the winner. They seem to be equal technologically; both have had victories and losses. Also, it's hard to decipher what connections exist with the other alien races—like who's helping who?

It's been reported that a race of extremely potent ETs, called the Blue Avians (Trevor's *Blue Bird* friends from the Sphere Being Alliance)—may be ready to step in and help the Nordics help us now that the Draco Reptilians have access to nukes. The Blue Avians are a superintelligent, benevolent race of ETs who purportedly help the Galactic Federation save endangered civilizations when all else has failed. They are spiritual beings from what is called the sixth density; traveling the galaxy in spherical, planet-sized spaceships. Recall the Empire's Death Star in Star Wars. Now, imagine it as a benevolent realty, a Life Star—many of them.

I've got my fingers crossed that Commander Trevor Smith and Colonel George Simon (the real men behind these pseudonyms) will get the *Blue Avians* to help the Bright Force.

The defeat of the planned nuclear attack on Hawaii in January of 2018 is a positive sign that the Bright Force can win out over the Dark Force. Maybe the *Blue Avians* have already altered the balance and are now helping the Nordics help us? As my wife would say, "Believe" . . ..

Author parasailing in Acapulco Bay —
the old way where you get some altitude

Chapter Eleven

Adventures In Acapulco
A Salute to Acapulco's Past

A*capulco, Mexico—October 1968*

Up here in the green hills, the early sunshine rapidly evaporates the dew, releasing that fleeting essence of a tropical morning. The iguanas intercept those first streaks of sunlight on the rocks behind the motel. They are big brutes, four-footers, always happy to devour whatever fruit I bring out for them. After a month they know me, but still alter their posture to confirm it's me—they're always wary of hawks. Beyond the lizards and two thousand feet below the hills, Acapulco greets the day.

My fraternity brother Tim Klein (T.K.) and I were on an extended holiday in Mexico before our scheduled induction into the United States Marines and an all-expense paid trip to a different tropical setting—Viet Nam. We had both graduated with majors in chemistry from the University of Wisconsin, Stevens Point (UWSP). T.K. is thin, fair complexioned, and quite perspicacious—a good friend.

We discovered a motel under construction in the hills, where the contractor was happy to rent us a partially finished room for fifty bucks a month. It was October 1968, and we had to be back in Wisconsin by December first, leaving fifty-five days of freedom before our lives would be reinvented.

We'd met a local man named Lopez at a beach restaurant a week earlier who swore he had taught John Wayne to scuba dive. After consuming several tequila martinis, each containing an embalmed worm, we were easily talked into purchasing a lesson.

The next morning T.K. and I were in our swimming suits, clipping along Acapulco Bay in a dive boat piloted by Lopez. We were accompanied by two Swiss couples and a local boy, Poncho, who was instructing us on scuba diving—like how to do it. None of us had any scuba experience.

"Spit in mask then rinse to prevent fogging. Blow bubbles through nose with head tilted back to clear water in mask." He demonstrated. "Don't come up faster than your smallest bubbles, make the OK sign (thumb with index finger) if you're okay, point UP if not; use snorkel and return to boat if you run out of air." Everyone seemed to understand the OK and UP signs and practiced sucking air out of their regulators. Poncho gave us a few additional pointers but it was obvious he believed in brevity. His yellow swimming suit complemented his tanned body and ebony hair.

I recalled a physical education instructor at the university who insisted that scuba diving required multiple lessons in a pool before going into a lake or the ocean. It was clear we'd be skipping all that . . ..

Lopez drove the boat past Playa Manzanilla beach then around the expansive Las Playas peninsula on the west end of Acapulco Bay, passing the Caleta Hotel and its Morning Beach on our right and Roqueta Island on the left. He pointed to a white house on a hill, claiming it belonged to The Duke. But before I could ask him about it, the anchor was pitched overboard, and we were told to put on our scuba tanks and weight belts. My mind was racing, anticipating the adventure. Being more discerning, T.K. was concerned about the velocity at which all of this was happening.

We could see a sea mount from the boat just off the bow—an underwater mountain visible just under the surface. Quite an assortment of multicolored fish surrounded it. The anchor held us off the mount in ninety feet of water.

I tested my fins in the boat, then switched to a tighter fitting pair. "Go down slow, hold onto anchor rope," Poncho instructed before he entered the water backwards with a splash. Lopez helped guide everyone else into the water except me. They all looked a bit frightened once in the water, holding onto the boat's ladder, each other, or the anchor rope. I was last, attempting the backwards entry with marginal success. The strap on my air tank wasn't tight enough, so my first scuba dive began with a head bump. Lopez gave me a stern look.

Suddenly, we were under investigation by a school of yellow-backed, black-striped sergeant majors checking us out for edible parasites. One of the Swiss ladies had freckles and, being mostly bare-skinned, became a target for the freckle-pecking fish. No blood, just firm pecks; the sergeants had no teeth. The Swiss lady, unamused, made a fast departure back to the boat with husband in tow.

Lopez grabbed my arm and started swimming down along the mount. I stayed with him, kicking hard. About twenty feet down the water cooled off and my ears felt the pressure of the depth. Lopez pointed to my nose. I held it closed and blew hard, like Poncho had instructed; the pressure equalized. Exotic reef fish swam all around the mount, but the challenges of my first dive preoccupied me.

We slowed down as we approached the thirty foot marker on the anchor rope; we were now one-atmosphere deep. Any deeper and the rules changed: you had to come up extra slow to prevent getting the bends. Lopez gave me the OK sign and I returned it. Looking up I could see T.K. and the other Swiss couple fifteen feet above us, looking at fish—Poncho was with them. Then I noticed the fish got bigger as we went deeper—a school of twenty-pound Crevalle jacks had just glided by.

Lopez looked at me, then pointed down. I nodded, pleased at how I was doing, and gave him the OK sign again. He motioned me to hold onto the anchor rope and ease my way down. It was obvious from my buoyancy that I needed more lead in my weight belt. I held my nose closed and blew it, again equalizing the pressure. When we reached the sixty-foot mark—two atmospheres deep—we stopped. It was colder and darker now, neither of us had worn a diving suit—just an air tank, weight belt, mask and fins.

Lopez smiled with approval as we kept moving down. We were seventy feet deep when suddenly a serious expression erupted on his face. He pointed down. *Shark!* I thought at first. Then I stared into the darkness of the depth and my eyes focused. *What the hell is that?*

Lopez tapped my shoulder and motioned for us to go up. I didn't need any encouragement. I used the anchor rope to go up slowly, allowing a burst of my smallest bubbles to rise six feet above me before proceeding to that spot and stopping. Then I repeated the process. Lopez stayed with me but kept looking down.

I couldn't take my eyes off what was below us, either. No shark, just a monster brown fish with an enormous mouth. It could have swallowed three of those twenty-pound Crevalle jacks in a single gulp. Two gold-circled black eyes the size of racquetballs peered from above its mouth, closely examining us.

My mind was racing; the only thing I could relate it to was an unbelievably huge large-mouth bass—one that weighed around eight hundred pounds!

Poncho had everyone back in the boat when we surfaced. T.K. and the Swiss couple who stayed with him hadn't gone any deeper than twenty feet.

"*Mero gigante de pescado!* Lopez yelled upon surfacing, pointing down.

"Giant brown fish!" I gurgled, raising my mask after surfacing. "Some kind of jewfish or world-record grouper."

"*Tres metros mero,*" Lopez asserted as he climbed into the boat. "*minimo quattrocento kilos.*"

"Three meter long grouper, minimum weight: four hundred kilos," the freckled lady translated as I climbed into the boat.

This started a discussion regarding large grouper fish, *gran miro.* I had seen a fat six-footer that had washed up on shore in the Florida Keys at spring break the previous year—a red tide victim. Called a jewfish in the Atlantic, it was huge but in distant second place compared to this one. We were later informed that we had encountered a rare goliath grouper, a step up from a giant grouper. Two other dive boats had reported it.

"Do they bite?" the freckled lady asked.

"No, they just eat you. Last year a diver almost lost a leg to one!" Poncho answered. "They have no fear."

"I saw one gulp down a four foot lemon shark," Lopez added while taking off his scuba tank. "A fisherman had caught the shark, and it was swimming in a circle alongside their boat. The huge grouper came up from deep water and swallowed the shark head first. Then disappeared."

"So much for worrying about sharks," T.K. remarked as we headed back to shore.

Later that day, T.K. and I were having a drink at an outdoor bar at the

Las Brisas Hotel.

"Exciting first scuba lesson," I remarked.

"There's nothing like a gin and tonic to settle your bowels down after their contents have been scared out of you," T.K. politely elaborated.

"I'll drink to that!" I agreed.

"I didn't feel comfortable going as deep as you did."

"If it weren't for Lopez taking me by the arm, I wouldn't have gone that deep on a first dive."

"We should do it again before leaving Acapulco. Need to go with a different group, though—one with better equipment," T.K. advised.

"Yes! Like BCs (Buoyancy Compensators) and wetsuits," I specified. T.K. nodded, then signaled the server to come over.

"Two more gin and tonics," I shouted. There was a mariachi band roving around the bar asserting their trumpets.

"Oh well, the U. S. Marine Corps will sort things out for us. I just don't want to be totally scuba ignorant when I get to basic training," T.K. hollered back.

"Me neither! I heard their final scuba exam is simple: They throw all your equipment into the bottom of a twenty-foot deep pool. You jump in naked, swim down, assemble yourself on the bottom, then swim up a fully functional scuba diver wearing all the gear: wetsuit, tank, weight belt, BC, mask, snorkel, fins, Mickey Mouse dive watch, etc."

"Wow," T.K. mumbled. "We better get more practice."

We did have good wheels in Acapulco. T.K.'s parents had let us borrow their 1966 Chevy GTO. We'd driven it here three weeks earlier from Steven's Point, Wisconsin—2,500 miles.

The Las Brisas Hotel was on Highway 200 near the east end of Acapulco Bay. It was famous for its pink Jeep rentals and our favorite "post-adventure" drinking spot. Our diving lesson had been on the far west side of the bay past the Hotel Boca Chica—another good place to have a drink that wasn't far from the famous Cliff Divers.

Our time in Acapulco was special. It went fast as we mixed diving and snorkeling with deep sea fishing, jungle hiking, cliff diving (yes, with the pros, but at fifteen feet, not 135 ft), partying with stewardesses, feeding birds and iguanas, and, I'm sure, some forgotten events. I didn't know then that Acapulco would become a frequent stop for me in future travels.

Acapulco—January 1983

My third visit to Acapulco was on the tail end of a business trip to Mexico with a tall, middle-aged Danish man named Hugo Jensen. Hugo was the marketing manager for the company I worked for in Milwaukee; I managed the company's microbiology laboratories. Our mission in Mexico City had been to instruct food manufacturers on techniques that reduced microbial contamination.

After a week of making sales calls and doing seminars in Mexico City, hampered by grueling traffic, jugglers and window washers at stoplights, slide projectors that didn't work, and Hugo getting sick from drinking the water, we were both ready for some R&R before heading home. I suggested Acapulco.

The Pacific Ocean was mostly calm that first night, but there were still some waves. It was rare to not have any. These were periodic waves—spaced apart with the ninth in a series being the strongest. So when it came to ensuring a safe depth in the "Gulch," the divers would stand on the cliff, swinging their arms to stay loose while counting. Then, when the ninth wave crested—assuring the Gulch was at least sixteen feet deep—they would jump. The rocks below them extended out into the water at the base of the cliff, so it was critical to jump out far enough. With arms straight out, backs arched, the famous La Quebrada Cliff Divers of Acapulco would dive into the sea 135 feet below. Four times the height of an Olympic high dive!

Hugo was mesmerized by it. We stood with a group of tourists on a triangular stone observation deck on the lower cliff opposite the diving cliff. The Gulch was in the middle of the water channel that came off the ocean between the two. We watched as a diver put his arms up, signaling he was ready. Then, when the sea was right, he vaulted into a perfect swan dive.

The crowd shouted and clapped when he surfaced in swirls of white bubbles in the center of the Gulch, smiling and waving.

"Bravo, bravo!" Hugo shouted. I was pleased to see some excitement in him.

A mariachi band on the deck of the El Mirador Hotel played "Guadalajara." More divers followed; some even did back flips.

For the big event, nine divers climbed the rocks up the cliff. They had to pick their way with all fours—hands and feet; the rocks were slippery. Floodlights and spotlights illuminated it all.

Once on top, 135 feet above the sea, the divers took turns praying at a small altar. Some had stopped at the one-hundred-foot mark, an alternate diving spot. The nine men would dive together holding torches in the grand finale for the evening.

The El Mirador Hotel's Restaurant was nestled in the crevice between the two cliffs. Guests sat around outdoor tables on its famous deck, which provided a superb view of the divers. It was where Elizabeth Taylor sat with Michael Todd on their honeymoon in 1957, and where Elvis Presley sang in the romantic movie *Fun in Acapulco* in 1963. But Elvis, because of security worries, sang in a recreated version of the El Mirador on a Hollywood set. Ms. Taylor got the real deal; Elvis never made it to Acapulco.

Once all nine divers, holding torches, were standing on the edge of the rocks they would dive from—having counted the waves, leg muscles cocked and ready to push off—they jumped simultaneously. As they soared into the air cameras flashed, recording nine perfect swan dives amidst an avalanche of sparks and fire. The crowd roared when all nine heads emerged from the sea, which was white with foam.

"My son was a diver here once; his name was Marco," An older Hispanic man said to me. He was standing on my right on the observation deck just after the show ended. The second "was" in the man's sentence gave me pause . . ..

"I'm Stanley. What's your name, amigo?" were the first words I could think of asking him.

"I'm Pedro." He reached out and we shook hands. Hugo then introduced himself and shook Pedro's hand, too.

"Marco died five years ago; he was just twenty-one," Pedro said in a solemn voice.

"I'm very sorry. Was it a diving accident?"

"No, he was murdered." I didn't know what to say . . ..

"The *Lugarenientes* make mistake," Pedro lamented. Lugarenientes arranged murders for the cartel. I knew that much.

"Marco was beautiful. My one boy." Looking very sad, he gave both Hugo and me a small photo of Marco. On the back was written: "I'm Marco—please remember me!"

Before Hugo and I could respond, Pedro walked away, waving goodbye. At the top of the stairs leading to the El Mirador, we saw him handing out more photos of Marco.

I looked at Hugo. "Let's go have a *cerveza*. I need to ponder this."

We were staying at the Hotel Acapulco Plaza on the Playa Del Morro. The easily recognized rock island—El Morro—was out in the bay to our left as Hugo and I stood on the beach after breakfast. Five hundred feet above it, a young woman was suspended under a parachute pulled by a motorboat.

"She's parasailing, amigo," one of the beach boys said to me. "You go next—just ten dollars!"

"Go ask that guy!" I pointed to Hugo, who immediately shook his head no and pointed back to me.

"Your friend wants you to go, amigo."

"No, gracias," I said.

I watched the young woman land standing up while running on the beach until the parachute deflated and the boat stopped. It looked easy. Then I saw Hugo give the boy ten dollars and point to me again.

I thought to myself, *I'm acrophobic, do I really want to do this?*

But, before I could argue myself out of going, the boy had strapped a life jacket on me. Too late! He then buckled me into the parachute harness. His instructions were simple: "Hold onto straps, start running toward ocean."

With no time to be nervous or scared, I was a hundred feet up in seconds and above the level of the tallest hotels in a few more. There was no falling sensation, just airflow grazing my suspended body as it moved forward. After the surprise of the first minute, I began to enjoy it—the air was calm and I was flying with my bare, hairy legs hanging straight down. A new perspective on them!

Hugo was the tiny stick-person waving below, presumably happy it was me, not him, doing this. The boat made a gradual turn east running behind El Morro Island while facing the Playa Condesa and its beach restaurants. Then as it turned back west, my legs swung out from the inertia of the turn—like on a swing ride at an amusement park. Paralleling the main city beaches now—I could feel the power of the boat's 150-horsepower motor pulling the parachute. Beyond Roqueta Island in the distance was the spot where I'd had my first scuba lesson fifteen years ago.

The rope that tied the boat to the parachute sagged in a big arc, and I could see all of Acapulco Bay and the charter boats outside it in the blue water—where marlin, mahi mahi, and record-size sailfish roamed. I estimated that roughly half of the fleet was out fishing, the rest were still docked at the city's downtown marina. This meant it would be easy to bargain a fair price for a full day of fishing. Hugo had never been deep sea fishing.

After twenty minutes, how to manage the landing on the beach entered my mind. The mammoth red and white parachute ballooned above and behind me and I could imagine how undesirable it would be to wind up hanging from some hotel balcony. I dropped that thought and focused on the beach as the boat turned perpendicular to it and slowed. The descent was smooth and quick. I landed standing up in the sand, running only a few steps before the parachute collapsed. Proud to have conquered acrophobia, I would do it again. Great fun. Hugo welcomed me back.

After the parasailing event we drove to the old charter boat marina on Avenida Miguel Aleman and found an abundance of available fishing boats. All were willing to bargain. I had done this before and knew to look around first, placing emphasis on equipment quality before getting into any negotiations. With Hugo walking behind me in a state of fascination, we strolled the dock.

"Hey amigos, go with us tomorrow. I give you good price!" the mate on the *Gran Pez* shouted.

"We guarantee sailfish or no pay," a man declared from *Tres Ami-*

gos—a forty-foot cabin cruiser.

"We're just taking a look," I told them as we walked by. Of course, that enlivened their sales pitches. I was most interested in checking their fishing gear, the boat was secondary. Penn International equipment was preferred. I wanted new model reels spooled with fresh monofilament line and sturdy rods with stainless steel roller guides. Hugo was getting an education.

Acapulco is one of the best places in the world to catch Pacific sailfish. They range in size from six to eleven feet and weigh between 80-220 pounds in the southern Pacific. They're the fastest fish in the ocean and can top *seventy miles an hour* when necessary. When caught, their colors intensify and they jump and dive incessantly, completely clearing the water.

Sailfish work together in teams, they race through large schools of smaller fish like ballyhoo and mackerel, slashing with their bills. As a strategically-positioned team they can't be outflanked by the school. Then, using their broad sail (dorsal fin) as a brake, they turn back to eat the ones they wounded. Pelagic, like marlin, they roam the oceans of the world.

We passed by the *Valeria Bonita* and the *Pez Aventura,* without much commotion. A guy wearing a captain's hat with scrambled eggs on the brim was having a siesta in the Valeria's fighting chair.

"*Hola amigos,* come see our rods and reels; they're the best." An elderly gentleman on the *El Pez Vela* (The Sailfish) was reading my mind. His new model Penn International reels and sturdy Fenwick rods boldly reflected in the sun. He waved us in, said his name was Francisco, and held out his hand to help me into the boat. Hugo followed. I checked out all six of the golden reels and gleaming white rods. "Big investment—*muchos pesos,*" I stated.

"Si, new this year," Francisco confirmed.

"Berkley's eighty-pound test?" I asked him, pointing to the line on the reels.

"Spooled three days ago, better than one-hundred-thirty pound for pez vela. More action," Francisco explained in clear English. This meant the bait would bounce and dive better when trolled. Eighty-pound line was considered heavy for sailfish in Florida, but here it was necessary. On good days, multiple charters fished in close proximity of each other and their lines could easily tangle. So, getting fish in quickly on heavier tackle was the rule. A 150-pound sailfish that could fight for two hours on twenty-pound test line could be brought to the boat in less than an hour on eighty-pound. There were many sailfish in and outside Acapulco Bay in January.

I nodded to Francisco, then checked the line-roller guides on the rods, all spun without resistance. A rusty or damaged roller could break the line—losing the sailfish along with many yards of monofilament line.

"Go with me. I promise two sailfish, maybe more." I paused and scratched my chin. I knew January was a great time to catch them in the bay, no long distance runs would be required before starting to fish. I could see from the parachute ride that charters had their lines out shortly after leaving the dock.

"How much for a full day?" I asked him. Francisco looked me in the eye, I could see wisdom in the man's face from years on the water.

"Three hundred dollars," he said. "No fish no pay." I knew of no charter captain that could make such a guarantee in U.S. waters.

"You don't have to guarantee fish; I'll guarantee them for you. I'll give you two hundred dollars now, and fifty dollars for every sailfish we catch tomorrow—no limit." This was not counterintuitive; I had learned from experience that keeping the charter crew turned on to catching fish for a full eight hours required inducement.

Francisco smiled; he understood the mission but had one question: "How many sailfish do you want to keep?"

"None, we must release them." I knew the locals smoked sailfish and ate them; they were not considered exceptional table fare.

Francisco smiled again, he explained it was common for him to keep sailfish for eating but he rarely kept more than two. They were plentiful in the Acapulco area and not endangered. Usually after two were caught on a guide date, he would change baits from ballyhoo to sardines and catch other fish. But he could tell that wouldn't go down well with me.

So Francisco agreed with my offer with one exception: he would keep just one sailfish for eating. I thought about it and discussed it with Hugo. He agreed, looking quite excited.

"I would very much like to do this fishing tomorrow," he declared with a certified Danish accent. Hugo was tall, forty something, and white as the driven snow—zero suntan. I looked like a local next to him.

I paid Francisco the two hundred dollars. As we shook hands, he reminded me to bring lunch and be here by 7:00 a.m. He would provide soft drinks and bottled water. Francisco then went over to Hugo who was now looking at pictures of big fish on the wall in the boat's cabin.

"You're from Denmark," Francisco said to him. "I recognize your accent. You will catch a big sailfish tomorrow." Hugo smiled.

Hugo and I walked the beach area's restaurants and bars that night. Hugo was enjoying every mariachi band in every place we passed, juking around, pointing his index fingers at other people as his shoulders moved with the rhythm of the music. I figured it was a Danish thing but couldn't recall anybody dancing like that in Copenhagen. I told him he had invented a new dance—The Sailfish Juke. I attempted to follow his moves.

We decided to eat at Poncho's Restaurant, where the mariachi music was played by four well-dressed, wide-hatted hombres. The bartender explained the two types of rhythm guitars the band used. The vihuela was the big round-backed guitar that produced high-pitch rhythm, supported by the guitarron, a bass guitar with its harmonizing lower pitch-rhythm. The high-energy music created a perfect atmosphere.

We both ordered the grilled, freshly caught black grouper accented with lime-coconut sauce—an edibility equivalent to a freshly-fried walleye at shore lunch on a Canadian lake.

Early the next morning we were underway, standing at the stern watching the hotels shrink on the Las Playas Peninsula as the boat moved out into Acapulco Bay.

Francisco steered from up top while Lorenzo, his son, rigged the lines with fresh ballyhoo—a medium-size baitfish with a 9/0 hook sewed into its belly with dental floss. This bait could catch both sailfish and marlin when trolled splashing on the surface. There were some secrets to getting a dead ballyhoo to splash properly, like breaking its back, cutting off its pectoral fins, poking its eyes out, and breaking off its bill. Just a few minor adjustments. Finally, a white and red assembly of marabou-stork feathers glued to a narrow wire cylinder was slid down the leader to the nose of the bait creating a wounded, bleeding look.

In short order, Lorenzo had six rods rigged and fishing—two on each outrigger pole and two straight off the stern. The ballyhoo skipped on the surface with an enticing rhythm. Hugo was transfixed, watching them.

"How fast are we going?" I shouted to Francisco.

"Nine knots (10.4 mph)," he shouted back. By Florida standards that was a bit fast.

"Not too fast for big sailfish," Lorenzo asserted. His English was as clear as his father's. He told me his father was a retired English teacher turned fisherman. I was feeling good about choosing them, very fish confident.

It was a perfect January day in Acapulco at eighty-two degrees with a mild breeze. I explained to Hugo how to use the Penn International reels as four pelicans flew over us eyeballing the splashing ballyhoo.

"Ever hook one?" I asked Lorenzo.

"Once," he replied. "When I reeled in, it went crazy—had to club it. Had no choice; it was either the squawking bird or me."

Francisco kicked the speed up to ten knots (11.5 mph) and the pelicans flew away, then he slowed back to nine. We trolled over water thirty-five meters deep (115 feet) just outside the bay for an hour.

Farther out, several other charter boats were trolling over deeper water, one had a sailfish on.

"Hugo, look quick, to your right!" I yelled. The sailfish had completely cleared water and was now horizontal, facing us in midair—between us and the other boat.

"I see it!" He said, just as the fish fell, flat-sided, back into the water, creating a massive splash—"like a horse falling off a cliff" to quote Hemingway. Then it jumped again. I could see It was a big sailfish—at least ten feet long, thick and big-headed.

"My cousin has it on!" Lorenzo declared. He could see his younger cousin, Ricardo, standing up and reeling. No guests appeared to be in the boat.

"That's the *Sea Devil,*" Francisco yelled down. "My brother's boat."

"My uncle is teaching Ricardo to fish," Lorenzo explained. "He wants to be a mate like me." Francisco was talking on the radio to his brother. The sailfish had been looking for a meal in over fifty meters of water when it ate their ballyhoo on the surface.

"Water temperature too high here, 30°C (86°F)," Francisco shouted down, "cooler where the fish are feeding." He headed out into deeper water; all six of our ballyhoo were on the surface, splashing perfectly. In twenty minutes we were over fifty meters of water (164 feet) with a surface temperature of 27°C (80.6°F). A dramatic change for such a short distance.

Lorenzo saw the first follow. The fish's broad iridescent sail was out of the water behind the ballyhoo on the close-in, port-side outrigger rod. It was slashing at the bait with its bill.

"Eat it, don't try to kill it—it's already dead," Lorenzo yelled to the fish.

"Get ready, Hugo!" I told him. He looked a bit petrified seeing an eight-foot fish so close.

Seconds later the fish ate the ballyhoo and the rod buckled in the gunnel; line raced off the reel. Instantly, Lorenzo pulled the rod out of its holder, pushed the reel's drag lever forward for more tension, pulled back on the rod twice to set the hook, then reduced the drag for the fight.

Lorenzo quickly inserted the butt end of the rod into the swivel cup on the fighting chair containing Hugo just as a second rod went off on the starboard outrigger. I grabbed it and duplicated Lorenzo's procedure, then shouted: "I'll handle this one Lorenzo, you help Hugo."

Hugo, white-faced, was holding the rod. "What should I do?" he blurted.

"Hold rod tight, don't let fish pull it away from you. Let fish run, don't wind reel yet." Hugo's sailfish broke water in a series of greyhound leaps as it bolted out to sea. My fish had hit farther out, we didn't see it follow.

"Two sailfish at one time!" Francisco shouted. "Don't let them tangle!" He had the boat moving forward slowly, turning as necessary to keep the two fish apart.

I eased up on the drag and let my fish get out farther, it was a big fish—maybe eleven feet, two hundred-pound class. Lorenzo had Hugo pumping and reeling now, attempting to get his fish in first. Here's where the heavy tackle was important. Francisco, standing up facing us, steered the boat with one hand behind his back so he could watch everything. If the lines tangled, we could lose both rods, imperil both fish, and possibly get hurt. We were not wearing shoulder harnesses, so being pulled overboard was not a worry.

I stood in the starboard corner of the transom and held my rod up, watching the amount of line on the reel. The fish was over three-hundred yards away and pulling hard. When it came up jumping, I could see it shaking its sharp bill. After the jumps, I cranked line in fast to pick up the slack. Slack line was a fisherman's enemy—that's when a big fish could depart without saying goodbye.

For thirty minutes the two fish had stayed apart, but then things got challenging. Hugo's fish was closer to the boat now. It was eight feet long and about 125 pounds. When it saw the boat, it veered my way and crossed under my line.

I ran with my rod to the ladder that led to the upper deck where Francisco was steering the boat. I had to go higher to keep my line over Hugo's. Lorenzo gave me a thumbs-up, he was at the transom and had Hugo keeping his fish low in the water, under my line. It was a close call. Lorenzo had his filet knife out, ready to cut both lines if they got tangled.

Francisco was down on the lower deck with us now. He had put the boat in neutral and was ready to help Lorenzo land Hugo's sailfish. Lorenzo, with rubber gloves on, grabbed the fifteen-foot leader and guided the fish close enough to grab its bill. Francisco positioned himself next to his son while holding a baseball bat, then waited for a few perilous moments while Lorenzo pulled up on the fish, its bill dangerously pointing at Francisco's heart. The ex-English teacher clubbed the fish twice. Their family would eat this one.

My fish was not as cooperative. I'd changed places with Hugo and was sitting in the fighting chair with a full bend in the rod. It took another hour of pumping and reeling to tire the fish enough to get it boat side. This was my largest sailfish at the time: ten and a half feet long and an estimated 210 pounds—a real beauty. Lorenzo cut the leader just outside its mouth; the single hook in the corner of the fish's mouth would eventually rust out. Once free, the great fish still had some joules in reserve. Electrified, it propelled itself out of the water just six feet from the boat—completely drenching us as it greyhounded out to sea. A fitting reprisal.

Hugo and I ate the Mexican versions of Cuban subs for lunch that we'd bought at Sanborn's in town, along with Inca Kola imported from Peru. We had the entire afternoon to catch more sailfish.

We caught four more that day, two each. They were between ninety and one hundred thirty-five pounds—we released them all. My arms and legs could tell it had been more than a day at the beach. Weightlifting coupled with arm and wrist isometrics had been at play. But, soreness notwithstanding, it had been a great day. Hugo had found another avocation to go along with his improvised mariachi dancing. I was happy for him.

On the way back to the marina, I sat up on the top deck with Francisco and noticed a photo of a young man clipped to the window. He looked familiar.

"Who's that young man?" I asked Francisco.

"That's Marco. He's no longer with us." I felt a chill go down my spine.

"Hugo and I met his father, Pedro, the other night by the Cliff divers. He was handing out photos of Marco."

"Pedro's been doing that for five years. He says it helps strengthen Marco's soul in the spirit world. He's my wife's brother."

"Pedro told us Marco was ordered killed by one of the cartel's lieutenants—by mistake."

"It was no mistake, it was intentional. They decapitated him; his head washed up on the Caleta Beach. He was my favorite nephew, never did any harm, no drugs, always happy with life."

I felt sick hearing this. "Why did they kill him?"

"Marco had many friends, some of them sold drugs for the cartel. He encouraged them to stop, a few took his advice. He was murdered as a warning to the other young drug sellers: *Stop selling drugs and your head will float with the fishes*."

"Is there any solution to this?" I asked Francisco. We were close to the marina now, passing the Las Playas Peninsula.

"Socialism is to blame; it does not provide profitable opportunities for young men. It punishes success, which leaves the door open for gangs and the cartel to employ them."

"It's a sickening, heinous story. When will society learn?" I sighed.

"The answer has to come from America, the United States. There has always been rich people and poor people. America created the middle class. It must fight hard to keep it strong and not let the fires of socialism change it into a collective of mediocrity. Socialism has failed the world—it has a losing record," Francisco articulated. "You can't redistribute wealth, it only spreads poverty.

"American's like yourself need to share your optimism and secrets of success with our young people—stay involved with us. At the same time Mexican elders, like me, have to be more forceful and demand a new generation of law enforcement officers that agree to pledge their souls to follow and enforce the laws. Mexico has the necessary laws in place to stop this evil—they just need to be enforced. I'm a conservative man, but one way to do this is with a 'law-enforcement tax' used to pay bonuses to those policemen and women who excel in enforcing the law. I'm serious, and I hate taxes! The cartel and gang leaders pay more for evil than we do for good!"

With that bit of wisdom said, Francisco backed into the *El Pez Vela's* berth at the dock. I put my hand on his shoulder and thanked him for a great day of fishing, and for his wisdom. I promised I'd pass it on. When I opened my wallet to pay him, a photo fell on the ground—it was his nephew's picture. I picked it up and showed it to Francisco and pledged: "I'll always remember him!"

Postscript

I wrote this story in November 2018. The memories of the events that happened three to four decades ago are as clear today as they were when I experienced them.

I have had the good fortune of visiting Acapulco in its glory days a total of six times. The days when Frank Sinatra sang "Come Fly with Me" from one of the hotel balconies. Today, I'm disheartened to say, it's a dangerous place. Because of the drug cartel and opposing gangs, it's now known as the murder capital of Mexico. This is tragic.

Tourists from the U.S. still go to Acapulco, but not as many. The U.S. State Department issues regular warnings. Seventeen hundred police are employed by the city of Acapulco, which averages over seven hundred murders a year. To put this in perspective, Chicago, with over four times the population of Acapulco, experiences five to six hundred murders per year, which I consider horrific.

Major hotels in Acapulco now employ drones to surveil their

property and armed guards to protect their guests. The socialist state of Guerrero, the home of Acapulco, has failed to provide opportunities and protection for its people.

Hopefully things will change in the future for this special place, and Americans will unite with Mexican elders like Francisco and whatever exemplary politicians there are to make this happen. Good must prevail—there's no acceptable option.

Paradise Tahitian Style — Moorea Island

Chapter Twelve

Confrontation in Tahiti

Big Shark . . . Small Boat

Prologue

I was standing on a rocky cliff at South Point, *Ka Lae,* on the Big Island of Hawaii in May of 2004. The windy day convinced me to tighten my Minnesota Twins cap or risk losing it. Ahead was a great vastness of open water—the South Pacific Ocean.

Stretching my arms out, one pointing southeast toward Easter Island, the other southwest toward New Zealand, each over four thousand miles away, I completed an equilateral triangle in my mind by drawing a line from New Zealand to Easter Island. Called *The Polynesian Triangle,* it enclosed over eight million square miles of Pacific Ocean, a sizable slice of the Pacific's 62.5 million square miles. Quite a formidable challenge for native people to conquer in canoes two thousand years ago.

They started out from Southeast Asia, what's presently Indonesia and Malaysia, and arrived in the western South Pacific around the time of Jesus, colonizing the islands of Vanuatu, Fiji, Tonga, and Samoa before heading further east, venturing fearlessly in what had to seem like an infinitude of blue water. At the same time, on the other side of the world, Greek, Spanish, and Portuguese sailing ships hugged the coastlines of the Mediterranean and Atlantic, afraid of falling off the flat earth if they ventured out too far.

Depending on which history books you read, the early Polynesians got to the Society Islands and Tahiti (French Polynesia) between 200 and 800 AD. Most of those who survived accomplished this in large, double-hulled wooden canoes propelled by sails that looked like crab claws. They navigated by the stars and the sea's secret instructions.

Almost due south of Hawaii, 2,600 miles, and somewhat below and slightly east of the Polynesian Triangle's center, is the island of Tahiti, a paradise graced by its remoteness, fern-green mountains, black-sand beaches, small deserted islets in turquoise lagoons, and an unforgettable fragrance.

The Society Islands are divided into two groups, the windward group of five islands including Tahiti and Moorea, and the leeward group of eight including Bora Bora and Raiatea. The prevailing southeast trade winds hit the windward islands first which are southeast of the leeward islands.

Moorea, April 2003

Moorea, one hour by ferry from Tahiti, was my primary destination for this trip. I had appointments at the Richard B. Gump Research Station run by the University of California, Berkley, to discuss the microbiology of coral reefs in the South Pacific.

After landing in Papeete, Tahiti on Air New Zealand, I rented a car and took the next ferry to Moorea. It was noon, local time, and I was hoping to find a fishing guide available for a half-day of arm exercise. From the ferry dock at Vaiare, I drove twenty-five kilometers to the Hotel Beachcomber Park Royal on the northwest corner of the island, parked, and hurried to the activities desk.

By 1:30 p.m. I was out in a seventeen-foot skiff with a young man named Hehu, trolling purple and yellow plastic squids in the Sea of the Moon—the water between Tahiti and Moorea. We had three rods out and rigged with the foot-long lures fluttering their tentacles near the surface as Hehu steered the 75-horsepower Evinrude. He was a muscular young man sporting a mop of curly black hair over calculating brown eyes and a Polynesian nose—a white cowrie shell necklace further certified his island heritage.

"My name means 'Rescued by God,'" Hehu declared. "You must

be God, Mr. Stan—you rescued me today."

"I'm not God, Hehu—just a scientist seeking adventure whenever I can. But thanks for the special compliment." It was a Monday, and he was happy to start the week making some money.

"What are our odds, being we're starting late?"

"Good for big sharks!"

"I bet they hang outside the lagoons where squid spawn."

"You know it!"

"What species?"

"Tigers, hammerheads, and lemon sharks." I could see an eight foot bang-stick suspended under the gunnel.

"Twelve-gage slugs?" I pointed to the bamboo pole with a shot-shell chamber at the tip.

"Or number-one buckshot—sixteen pellets. Either one. Depends on type and size of shark." I could already tell Hehu was the right guide to choose.

"You can be a tad less accurate with buckshot," I remarked. Hehu smiled. When an angler brought a shark close to the boat the mate would bang it in the head with the end of the bang-stick, detonating the cartridge. Once in the boat, the only good shark was a dead shark.

"Are they legal in the U.S.?"

"Bang-sticks are legal for hunting alligators in Florida, but can only be used for self-defense with sharks."

"How does a warden know if you killed a shark in self-defense?" Hehu questioned.

"It's up to the fisherman to convince him!" I grinned as I tightened my cap.

Hehu grinned back and followed the coastline of Moorea trolling at six miles an hour.

I took my shirt off, put on SPF-80 lotion, grabbed a seat cushion for a pillow, laid down—feet stretched over the side—and fell asleep. Jetlag! It had been a nine hour flight from LAX.

It was 4:00 p.m. when the boat jerked suddenly and woke me up. The starboard rod was bucking violently, and a fish was taking out line. By the time it took for me to sit up straight, Hehu had set the hook and was handing me the rod.

"Big tiger!" He declared.

His three rigs had Penn International reels spooled with 130-pound test monofilament. The line raced out against the drag. I stood up, held the rod high, and let the shark run. Hehu got the other two rods in quickly. We had to duck and maneuver around each other to keep from getting tangled—small boat.

We were outside Moorea's north lagoon which was full of four to five-foot black-tip sharks—regulars that hung around for the "shark feeding" snorkeling tours. When my son Jeff and I were here in 2000, black-tips were all over the place, along with scores of stingrays. It was unnerving at first having them all around us in the shallow water, but guides had trained them to eat pieces of squid which, purportedly, they preferred over pieces of people. Jeff made sure we always had enough squid.

When the big shark slowed down, I was able to crank in line. "It doesn't feel that big! Did you get a good look at it?" I asked Hehu. I'd caught a lot of five-to-six foot sharks when trolling for marlin in Mexico and the Caribbean, so I had a feel for them.

"She's close to twelve feet! Saw her engulf the lure—big female with huge head!"

"Seriously? Holy Macaroni!" I kept the rod up and tightened the drag a touch. The Penn International reels have a lever drag that's kept forward and tight while trolling to help set the hook, then pulled back to less tension for the fight.

"She's pulling harder now . . . seems to be swimming in a deep circle." I was standing up in the middle of the skiff on the flat part of the floor. I had the butt- end of the rod in a rod-cup around my waist—fully awake after hearing her size.

"*A twelve foot tiger shark attached to me in a seventeen foot boat.*" The thought took control of my mind. During World War II shipwrecked sailors in life rafts were often attacked by tiger sharks—usually in the dark.

"She doesn't know she's caught yet!" Hehu watched the angle of my line. "She's used to eating big things that kick and bite. She'll come up soon to investigate, to see what she ate. Be ready. She won't be happy to see us!"

The shark was bulldogging, pulling left, then right . . . staying deep, shaking her head. With each shake about six feet of line zipped off the reel. "I can feel her now; she's right below us." I could see concern in Hehu's eyes.

"A twelve-foot female tiger shark will weigh her length in units of a hundred pounds per foot—this one is twelve hundred pounds!" Hehu declared. I knew what he was thinking: *a mad tiger shark, green from just being caught, could tip us!*

She kept deep with a heavy, steady pull. When I pulled back with the rod, the reel gave line. There was no budging this monster shark against her will—I knew then that she was big. I also knew the 9/0 hook on the plastic squid was just a nuisance for her—a pin prick. She would decide when to come up. I sat down and put the rod between my legs, giving them a rest. An hour passed—she was pulling us out to sea. It was 5:10 p.m.—serious isometric time for the arm and back muscles. The secret method now was just to hold onto the rod.

I took a few minutes to look at the island of Moorea, arguably the most beautiful of the thirteen Society Islands. Her spiked, verdant green mountains merged with her golden beaches and turquoise shallows. It was enough to have convinced Charles Darwin that his theories of coral-reef evolution were correct. While looking at Moorea from a hilltop in Tahiti, he called it, "A picture in a frame." The mountainous, fifty-two-square-mile island was perfectly framed by the turquoise lagoons surrounding it.

"What's your biggest fish?" Hehu asked me as I pulled up on the rod, let it go down, then cranked in line from four revolutions of the reel's spool before lifting the rod up again. "This is my secret pumping technique," I joked. Hehu laughed.

"Seven-hundred-pound blue marlin." I took a breath.

"Where was that?"

"Off Bimini in the Bahamas."

"The other side of the world for me," Hehu said.

"Lost an even bigger fish, a nine-hundred-pound black marlin, off Cairns, Australia two years later."

"So this tiger shark could be your biggest fish!" Hehu looked pleased.

"If we get her in. She's coming up now!" I reeled fast, guiding line back on the reel with my left thumb.

"Give her some slack!"

"Understood!" I acknowledged and reeled slower—just the opposite of what to do with most fish. We didn't need a super-tight connection if she surfaced close and lunged at us. The draft on this skiff was only two feet.

When she broke water twenty feet astern, I was immediately stunned by her size and royal purple sheen. Many dozens of teeth angled sideways in her mouth, improving her ability to hold and eat hard objects, like big turtles. She knew at that moment we weren't something to eat—but something to hate. *Could sharks hate?*

Flee-versus-fight hormone activity must have been flowing inside her. In what seemed like a couple minutes, but was scarcely a few seconds, she turned, jerked hard on the rod and brought me to my knees—then bolted away at a velocity equivalent to any marlin I'd ever hooked. Line screamed off the reel as I held the rod high while sitting on the floor of the skiff.

I fought the tiger shark for another hour, getting as much of the 130-pound line back as possible. She was methodically pulling the boat and staying deep. Hehu had a swivel cup clamped onto one of the ribs on the floor of the skiff. I sat on a seat cushion on the floor and held the rod, butt end in the floor cup. This allowed me to spread my legs and get leverage. Unlike a Mako shark that can race and jump like a marlin, this giant tiger shark was in no hurry. It simply kept deep with a steady pull—*how long will it take her to tire and come up again,* I wondered.

"Hehu, how big can female tiger sharks get?" I grinned to myself at the alliteration possibilities with his name: Heeehuuu, Heehuuu; Hehu-Hehu-Hehu, and so on . . .. Serious fishing can do funny things to a man's mind.

"Sixteen-foot females caught off Australia by Japanese trawlers have exceeded two thousand pounds. I believe one eighteen-foot, pregnant female weighed over three thousand." Hehu started the motor briefly, then slowly corrected our position so the fish was pulling straight off the bow.

"I know it's an apex predator with a reputation for eating anything: license plates, tires, plastic toys, oil cans, fishermen!" I explained, just before the rod lunged forward powerfully. I tightened my grip and cranked in twenty feet of line before it went taut again. "What the hell was that?"

"She probably turned and ate something big, unfazed by the foot-long lure in her mouth." We kept discussing tiger sharks . . ..

It was 6:10 p.m., a half hour past sunset and into the twilight time, when Hehu and I had agreed that our mission was to get as much of the 130-pound line in as possible, then cut it. We hated to do this but it was sensible. Trying to unhook a giant shark next to the boat in the dark would have redefined stupidity—being my biggest fish notwithstanding. The 9/0 hook was wired to a stainless steel leader that was tied to the monofilament line. We would cut the line well back from the leader; the hook would rust out in time and take the leader with it. An annoyance for the shark until then, but we had no choice.

I jerked on the rod, goofing around, testing to see if the shark would respond. Nothing. I couldn't alter the tiger's deep, pulling force. It made me think of all the big fish I've been fortunate to catch. Twenty-five years ago, fishing guides would tell me they'd trade all the big fish they'd caught for the ones they lost. I figured at the time it was fictional guide talk, meant to impress clients. It wasn't; I could honestly claim the same now.

Suddenly, the line went totally limp. *She just threw the lure,* was my first thought. I stood up and lifted the rod with little resistance. Some line floated on the surface.

"Gone?" Hehu shrugged. I lifted the rod again while reeling slowly. Nothing. *Did she race toward the surface super-fast and beat the line?*

Whoosh! Whoosh! "She's up—ten yards off the bow," I shouted. "And still on!"

"I see her," Hehu yelled as he reached for the bang-stick. Then he held it over the side of the skiff—ready to bang her if she charged. We were on the dim side of the twilight zone—it got dark fast in the tropics. Our strategy had changed. We had to be ready to defend ourselves.

"She's right here," I barked, and ran forward reeling in slack line until loops of it jammed the reel. My left thumb had screwed up and didn't guide the line evenly! The shark was right at the bow when the line tightened—she'd rolled up in it and couldn't swim. My mind raced. I've had fifty-pound muskies do this in Canada, but it was unimaginable with a 1,200 pound shark. *What should I do next?*

Hehu composed himself and held the bang-stick alongside me. I was kneeling on the floor near the bow. The shark, making jerking reflexes, pectoral fins held tight against her body by a dozen wraps of line, slithered like an anaconda—rubbing against the starboard side of the skiff, tilting us. Hehu rushed to put his weight portside and so did I, crawling on the floor while holding the rod up with one hand.

I stood up and put the rod's butt back in the cup strapped around my waist. Twelve hundred pounds of slithering, squirming tiger shark was jolting me. *How long would the wraps of line control her?*

"I can bang her. She's not moving fast," Hehu shouted. He could see how well she was rolled up in the line. Then we'll tie her to the skiff and bring her in . . . hang her from the light pole by your overwater bungalow at the Park Royal. She'll dwarf you! You'll be famous in the morning—an American hero!"

I put what Hehu had just said out of my mind. *My God, she's squirming her way around the skiff.* I could see the plastic squid in her mouth while I followed her, rod held high with tangled line everywhere. I was amazed we were still connected and that I could guide her with the rod, realizing my legs could easily get tangled in the gobs of line that were hanging loose off the reel. I couldn't tell what was loose from what was tight. She could easily deep-six me, and that would be all she wrote.

"What an unbelievable monster shark, twice my height in length and easily six-feet in girth!" Her swept-back tail fin was five feet above the water. Hehu was ready with the bang-stick.

"No!" I yelled. "Hand me your long gaff." He looked confused when he handed it to me. I knelt where I expected the shark's head to be in a few seconds. Then—leaning over the side, two feet from the ferocious mouth—I thrust the gaff's hook around the squid lure and twisted the handle to catch the leader, pulled back hard, and unhooked the shark. Still rolled up in the line, she could now unwind. Hehu started cutting all the loose line that was wrapped around my feet. The rod, reel and fisherman were secure. Whew!

"Let's get our arses going!" I bellowed.

"I'm starting the motor, hold on." Hehu turned the boat and put some quick distance between us and the shark.

Both fish and fisherman had won! But winning had required me to do something that redefined stupidity: Unhooking a 1,200 pound tiger shark in the dark from a small boat. Sometimes in life, you get the chance to eat your own words. This was one of those times. However, there is often a thin line between being a hero and being a fool—in this case, the monofilament was that thin line.

We headed toward Moorea, breathless. We could see the tiger shark spinning herself loose as we passed a large squad of phosphorescent squid, activated by our wake. "Big tiger sharks eat squid!" Hehu shouted over the motor rattle. I managed a grin.

"You are a great fisherman!" Hehu held his thumb up.

"Thank you!" I exclaimed back. Heart pounding, I shook my head in relief.

Hehu steered southeast toward the Sea of the Moon, between Tahiti and Moorea. The older I get, the easier it seems for me to calm down after a risky adventure. If I were a cat, I'd certainly be in my ninth life. While on the flight from LAX to Tahiti, I'd assumed I would just take it easy on the first day. Stay around the hotel pool, have a couple Hinano beers, sit on the deck of my overwater bungalow, and watch the resident pod of bottlenose dolphins enjoying life, maybe even kick back and take a mid-afternoon nap. But, after landing and seeing the perfect day, with at least half of it left, my inner voice told me to go fishing. I've learned to listen to that voice.

Moorea's sharp mountain pinnacles penetrated the night sky, creating jagged shapes with only two dimensions, like giant cardboard cutouts held up in the blackness—shapes that hid some of the southern hemisphere's emerging stars. We slowed down as we neared Moorea's coral reef. There were twelve safe areas to pass through—you had to know at least one of them.

Realizing we both needed some calm time, Hehu slowed down and pointed up at the stars.

"Look, you can see *Epsilon Pavonis* rising on the horizon—a bit west of south." Hehu pointed southwest. "It's in the constellation *Pavo* (peacock), one of the first to show."

"I can see it, just a few degrees above the horizon." I pulled the brim of my cap up to see it better.

"At midnight it will be ten degrees above the horizon," Hehu said.

"Does it ever set?" I knew Tahiti was at 17 ½ degrees south latitude and stars further south usually just circled the South Pole.

"Comes close within one degree but never completely sets. Most stars rise in the east and set in the west as the earth turns. My ancestors depended on at least thirty stars for navigation. We can see about two thousand with the naked eye."

It was amazing . . . the stars were calming me down. Like magic, my pulse had normalized.

"Throughout the night, my ancestors would watch those important stars when they first appeared above the horizon in the east, then as they proceeded to the zenith and ultimately set on the western horizon as the night advanced. They would correlate their east-west observations with those from north-south constellations to get a bearing.

"Examples of important stars the ancient Polynesians followed this way: Rigel, Spica, Sirius, Antares, Fomalhaut, Betelguese, Regulus, Pollux, Castor, Capella, and Vega. These eleven stars had to be completely memorized, where and when they rose in the east, where they were in transit across the sky, and where they set in the west. This information changed slightly every night!" Hehu elaborated. I could tell he was proud of what he knew.

Hehu then pointed out some of the well-known northern constellations that could still be seen, like the Big Dipper and Cassiopeia. "Notice anything unusual? He asked me.

"They're upside down! Cassiopeia's big 'W' is a big 'M' and the Big Dipper is spilling instead of scooping."

"Correct!" Then he pointed out the constellation Scorpius: "The reddish star Antares is at the scorpion's heart. Above it, the two pincers flare out. The tail is long and curves into a hook at its end, piercing the Milky Way."

"And the Southern Cross? *Crux*, my favorite southern constellation." Hehu pointed south, a hand's width above the horizon.

"Not so bright tonight. See the two pointer stars, Alpha and Beta Centauri?" I looked at his compass to get a bearing.

"I see them."

"They point up to the star, *Gacrux*, at the top of the Southern Cross."

"I see the Cross now. It's hazy, the bluish brightness of the star Acrux gives it away," I knew it was the thirteenth brightest star in the sky and at the bottom of the cross.

"Now, imagine a line from Gacrux to Acrux that extends out four and a half times the distance between them."

"Okay, now I'm with the program: Straight down from the end of that line is due south." I was proud that I remembered.

"You got it." Hehu smiled.

"I recall the Southern Cross, higher in the sky in Australia and East Africa," I could hear Crosby, Stills, and Nash singing about it in my head. "What happens when clouds roll in? How did your ancestors find their way without the stars or a compass?"

"They learned to read the sea, its currents, swells, and creatures. Swells generated by waves from distant winds can pulse the open ocean for thousands of miles. Down here the southeast trade winds generate swells. If your canoe is rolling sideways, it's running perpendicular to them; if its pitching bow up, it's running into them. Swells interrupted by an island form angle patterns in the water that can predict the island's location. Patches of seaweed follow currents that run in predictable directions. Birds fly out in the morning to feed and back to their home island in the evening."

"You're a smart young man, Hehu." I was totally calm now—joyous that the giant shark was swimming free in the Polynesian Triangle, and that Hehu had advanced my navigation knowledge.

"My father knows the sky and sea better than me. He has memorized all the important stars and the paths they take as the earth turns." Hehu goosed the RPMs on the Evinrude. After entering the lagoon, we were dosed with the sweet fragrance of frangipani flowers. The Beachcomber Park Royal was up ahead.

I was starved. I had shared a Snickers bar with Hehu, who shared a papaya with me; that was it for lunch. Air New Zealand had provided dinner the previous night—some type (unknown to the flight attendants) of baked fish with a tablespoon of peas and a miniature salad. I required significant nourishment after the shark fight. Hehu wished me well, he had chores to do at home—like changing diapers.

We would see each other again before week's end.

I drove to a small French restaurant called Le Rudy's—the place was recommended in the travel guide. Given my probable low blood-sugar, I didn't need any more convincing than that. A lovely Polynesian girl escorted me to a table by a window, looking out over the coastal road that circumnavigated the island. I requested a glass of the house Beaujolais—just like I would in Paris—did a quick scan of the menu, then ordered a medium-rare, New-Zealand filet mignon with morels and red yams.

Just as I was diving into a basket of warm French bread, an older lady at the table next to mine spoke up. "Excuse me, young man. Ellen, Mary, and I wonder if you know who played Fletcher Christian in the first two *Mutiny on the Bounty* movies? She looked at the two ladies sitting across from her who smiled and nodded.

"I think Clark Gable was in one of them," I suggested.

"No, he was in the third one," she assured me.

"Somebody told us there were five versions of the movie!" the tallest of the three ladies interjected.

"Never knew there were five!" I said, with a mouth full of bread. "Let me make sure I've got our names straight. I'm Stan."

"I'm so sorry, didn't mean to be rude. Ellen's the tall one, Mary's shorter and broader, and I'm Joyce. We're retired schoolteachers from Banbury, England."

"I have friends outside Banbury, they live in Hinton in the Hedges; Pat and Doris Finnegan," I said.

"You're kidding us!" Joyce replied. "We know Doris, we're in the same book club."

"What are the odds of that?" It gave me pause. "Banbury's close to ten thousand miles from here."

"There are no coincidences," Mary added. Her comment took us off course for a while in the direction of the spirit world. The ladies were having a late tea and could tell I was quite hungry by the way I examined each revolution of the kitchen door. Switching to a discussion on fishing tiger sharks would have been too much of a diversion. Besides, I was curious, hearing there were five Bounty mutinies in addition to the real, authentic mutiny in 1788.

"Getting back to Fletcher Christian, I know Mel Gibson played Fletcher in the 1984 version called *The Bounty.* Marlon Brando was

Fletcher before Mel, sometime in the 1960s, Gable in the mid-30s—those two movies were titled *Mutiny on the Bounty*." I forgot I'd remembered that. "And all three were filmed onsite here in Moorea in Opunohu Bay—close to where the Beachcomber Park Royal is now."

"Yes, we know that," Joyce assured me. "And Anthony Hopkins was the best Captain Bligh of them all—replacing wickedness with caring after being sent to sea in a small boat with his loyalists, by Fletcher (Mel Gibson)."

"Marlon Brando fell in love with his beautiful co-star, Tarita Tetiaroa, and married her in real life," Ellen explained after taking a bite of a frosted muffin.

"That's right, then he bought an island down here, some atoll, and raised a family on it," Joyce said. "He loved these islands."

"And they still live on them! Marlon refused to leave in 1973 to come get his Oscar for winning best actor in *The Godfather*," Mary added.

"And don't forget, Mel Gibson fell in love with Tivaite Vernette who was his Tahitian wife Mauatua in the movie—she was filmed topless with him!" Ellen elaborated, smiling at Mary.

"She's a celebrity down here now, travels around the Society Islands—was in Bora Bora last week." Mary pulled her chair in, adjusting herself.

"I'm sure Clark Gable fooled around, too," Joyce interjected. "They all did!"

"This tells me there were bonuses for running breadfruit pods from Tahiti to Jamaica (England intended to grow breadfruit in the Caribbean in the 1700s.). Too bad a mutiny got in the way," I remarked, just as my steak arrived.

"Actually, most of the real mutineers wound up on Pitcairn Island, truly in the middle of nowhere, with their wives and girlfriends. So the bonus program continued," Ellen assured us. "Art imitated real life the best in *The Bounty* version. But I can't speak for the first two movies."

Before the server left, I asked him if he knew who starred in the first two Bounty movies. He shook his head and said: "I'll ask Rudy, he'll know."

Shortly, a sturdy man in white chef's attire came out of the kitchen and walked over to our tables. Rudy enlightened us: "The first two movies were filmed in Australia, the first was a silent movie made in 1916—don't know who was in it. The second was called *In the Wake of the Bounty* and starred Errol Flynn in his film debut in 1933."

"Errol Flynn!" Joyce proclaimed. "Now we know. Thank you, Rudy!"

I winked at her while chewing my steak.

As I drove back to the Beachcomber Park Royal and my overwater bungalow, I wondered if any of the Fletchers had ever caught a decent size fish—maybe Marlon did? He had the name for it.

Cook's Bay—Moorea

The morning came with a beautiful sunrise breaking out over the mountains that guarded Opunohu Bay. The view was breathtaking and matched sunrise scenes in *The Bounty.* I was breakfasted, coffeed-out, and dressed in my swimming suit, flip flops, and Minnesota Twins tee shirt. I got in the Toyota rental car and headed for a morning of snorkeling in Cook's Bay.

The Gump Research Station was close by, and I had a change of clothes in the car for my meeting. I could morph into tropical business attire in minutes.

I'd brought my own snorkel gear—always did when travelling in the tropics. Rental gear is synonymous with loose fins, leaky masks, and snorkels that flood in the waves. I like to make sure the adventure part of any snorkeling event had nothing to do with surviving faulty equipment.

The clearest water in the world is in the South Pacific. Here in Tahiti and Moorea you don't need a boat; you can easily walk out into any turquoise bay or lagoon from the beach and start snorkeling. I sat down in two feet of water, put on my fins, ankle knife, mask cleared of spit, and vortex snorkel. Then, lying flat, face down, arms at my side, I headed out to the reef making as little turbulence as possible. I call it stealth snorkeling.

But, before I went out deeper to the coral banks—alive with an astonishing variety of reef fish—there were three types of critters I needed to remind myself to be particularly watchful for: Stonefish, Cone snails, and Lionfish. First timers should take a brief refresher course on these before jumping in [1], [2], [3].

When I arrived at the coral humps, I noticed I was being followed. Three bottle- nose dolphins were behind me. Looking straight down and watching the bottom for rays and stone fish, I didn't notice them. I could tell they wanted to play, but I hadn't brought a volleyball! I figured they were probably trained by scientists at the research station.

Coming into the reef from the shallow side, I was immediately surrounded by butterflyfish—exactly what I wanted to see. They feed on coral polyps almost exclusively. There at least 114 species of these multicolored, pulchritudinous fish which are about the size of a coffee-cup saucer and ubiquitous throughout the Pacific and Indian oceans. Some species are unique to a specific area while others are commonly found on numerous reefs, thousands of miles apart. To

1 Stonefish: They are the most poisonous fish in the sea and look like an encrusted rock or lump of coral—mostly brown or gray, sometimes with red or yellow spots. They like to hide, buried in sand on the bottom. If you step on one, its venomous dorsal spikes will inject you with a potent neurotoxin. Excruciating pain will confirm you've been envenomed! This is a medical emergency and you need antivenom ASAP—make sure you know in advance where to get it (clinic or hospital). Hot water helps destroy the venom, and should be used as first aid. Vinegar reduces pain. Don't use a compression bandage or remove the stingers. Antivenom is a must—stonefish venom can be lethal.

2 Cone snails: All produce conotoxins and the larger ones can kill you—some have shells up to nine inches long. There are over 3,000 species in the Conus genus—all are gastropods (snails) armed with poisonous harpoons used to spear fish, or humans that pick them up. Their shells are quite attractive and collectable (minus the snail). Beware of live ones! One large species called the Geography snail has the nickname "cigarette snail." After being harpooned (stung) by one, the victim only has time to smoke a cigarette—a bit of gallows humor. Another one called the textile cone has an orange conical shell covered with black stripes, interlocked with white "mountains"—some say the pattern resembles a Turkish rug. It's also quite dangerous. Cone snails can't swim so they're always found crawling on the bottom or on coral humps. Don't pick one up!

3 Lionfish: Unlike stonefish and cone snails, these guys don't crawl around on the bottom or hide in the sand. They swim around the reefs. You can't miss them! They flaunt conspicuous, feathery pectoral fins and up to eighteen venomous spines that are only used for self-defense. Grab one, and you'll get stung. It won't kill you, but it will spoil your day. It feels like getting stung by a magnum parasitoid wasp in the Amazon. I've experienced both: I backed into a lionfish while filming a fifty pound parrot fish in the Philippines, and the wasp got me on the shoulder in Manaus.

the extent that multiple species populate a particular coral reef system—more species indicating greater biodiversity—the health status of the system can be assessed.

For example, the Tahitian butterflyfish—brown, yellow-tailed, with a black stripe through its eyes—is unique to the Central South Pacific, while the gaudier pyramid butterflyfish seems limited to Hawaiian waters. The raccoon, ornate, and threadfin species appear equally throughout the greater pacific. At least that's been my observation.

I sleuthed over to a dozen double-saddled butterflyfish—white and yellow with two black vertical bars (saddles). They were giving coral polyps their undivided attention—swimming with precision while eyeballing them, looking for those with the worm exposed. Then—hovering—they would dart at a worm, brake with their pectoral fins, and nab it before it retracted into its tube. A sprint is equal to three darts for a butterflyfish; it can reverse in an instant if necessary. Soon, several threadfin species joined the party, there was a plethora of polyps on the healthy coral. I was right above the action in ten feet of water.

Moving on, I was pleased to spot a small, green sea turtle about eighteen inches long. It was not afraid of me, probably acclimated to people from inviting itself on the shark and ray feeding tours. Several blue, five-striped, pacific sergeant fish were pecking parasites off the turtle's carapace, which it seemed to enjoy. Green turtles don't smile, they wear permanently stern expressions—reminiscent of the nuns who taught Sunday school in the 1960s.

Below the turtle and near a large outcropping of coral, a titan trigger fish was moving chunks of coral around, looking for lunch: crustaceans, mollusks, and echinoderms were usually on the menu. Following his dusty trail, a Moorish Idol checked for leftovers. I watched the trigger fish carefully, making sure I stayed out of his territorial domain; titan males aggressively defend a conical zone that extends to the surface. Don't let his dazzling colors fool you, he can rise to the surface and bite if you inadvertently swim through his cone. While not poisonous, its bite feels like one from a small parrot. Whether the triggerfish bites or not depends on his mood—which is usually ill-tempered. Shaped like an oval dinner plate, he did manage to lift his conspicuous eyelids to get a full view of me. Fortunately, a stream of starfish larvae—free swimming for another week or two before sprouting arms—came out from under a rock and got his full attention. Lunch was served.

There are forty species of trigger fish, and not one has a dull complexion. Their color patterns are so spectacular and unreal that they seem alien to our world. And they're smart fish! Hawaiian university studies have shown they learn from experience—they can remember how to exit a Plexiglas maze. Males and females prepare for spawning by blowing sand on their planned egg site, while touching each other as practice for spawning. Triggerfish foreplay! During actual spawning they cooperate and exhibit bi-parental care. I gave the titan triggerfish a little more clearance and snorkeled on . . ..

Next, I was treated to two completely blue Crevalle Jacks—ten pounders—chasing after a school of spotted wrasse. I've seen them in Maui too, sleek like their grey cousins but bluer than the sky. Opposite them was a spotted eagle ray gliding through the water, black with yellow spots on top of its wings, pure white underneath. Poetry in motion. I stayed clear of its six-foot tail.

There are many unique fish in these cherished Society Islands—one special archipelago of many in the Polynesian Triangle.

Postscript

The Gump Research Station is located in Cook's Bay on Moorea Island. Thirty three acres on the bay were donated by Richard B. Gump in 1985. The station is managed by the University of California and is dedicated to studying the effects of climate changes, tropical biocomplexity, coral reef sustainability, molecular biology and more. This was my second visit to the Gump Station.

The facility is located in paradise, literally. I've been close to everywhere on this planet, and this research station gets six stars out of five. Facilities include wet labs, a flow-through seawater system, a molecular laboratory, general use labs, docks, research boats, conference room, offices, a fleet of vehicles, and a spectacular view of the bay guarded by several of Moorea's vivid mountains.

The project I was helping with involved stimulation of larger species of zooplankton that ranged in size from five to twenty millimeters. Examples are the decapods and mysids—small shrimplike crustaceans. The objective was to provide feeding options for fish—like the butterflyfish—that were dependent on coral polyps. Overgrazing polyps puts stress on the coral community, which is already stressed by changing climate and other factors.

Bigger, nutritious zooplankton feed on smaller zooplankton like copepods which feed on bacteria. Certain gram-positive bacteria in the Bacillus genus are particularly suited to improving the zooplankton food chain. My company's experience with prawn aquaculture in Southeast Asia was presented as a model for how this might work in coral-reef ecosystems. The challenge was how to translate our success in a closed system (a shrimp pond) to an open system (a coral reef). Bacillus bacteria easily wash away in open systems. So, our strategy was to incorporate natural biofilter materials which allowed Bacillus species to adhere, colonize, and continually inoculate the water with probiotic Bacillus. At the point we were at in 2004, this was just a theory. Plans were being discussed to test the theory in a seawater flow-through pilot system before taking it out to one of the lagoons. The process is now in use in Southeast Asia.

Confrontation in Tahiti

Charter Boat Row in Key West

Chapter Thirteen

Tarpon Times in Key West
Old Fishermen Never Die

Prologue

The tarpon, *Megalops atlanticus,* also called the Silver King, has been around for 125 million years. It's a superb game fish to catch on spinning tackle in the ocean.

Megalops (meaning "big face") does well in warm, oxygen-deficient water by gulping air and storing it for breathing in a unique air bladder. Individual fish can live for decades; one captured tarpon was sixty-three years old. The all-tackle world record weighed seven ounces shy of 287 pounds and was caught in 2003 off Guinea-Bassau on the African coast. The largest tarpon recorded was over nine feet long and weighed 355 pounds.

You don't eat tarpon, you release them. The objective is to have fun reeling in a giant fish on spinning tackle while it does aerial acrobatics and tries to pull your arms off. You do this standing up with your fishing rod bent in half, fighting the fish without any help. Then, when you convince it to come to the boat—an hour or two later—you unhook it in the water and set it free. Your mission is to make sure neither you nor the fish get hurt—just tired. My favorite place to fish big tarpon is Key West Harbor, from March through April.

Key West, Florida 2004

I was drinking a *Cuba Libre* at Sloppy Joe's bar on a warm March afternoon. The female bartender made it with dark Cuban rum, Coca Cola and a fresh-squeezed Key lime. I asked her to make a double in a beer stein. The extra volume made it easier to get the ingredient ratios correct and the stein was easy to hold. A few of my scientist buddies disputed that reasoning.

The bartender's name was Sally, an attractive young woman on roller skates who wore a tee shirt with Ernest Hemingway's face on it. Ernest was indisputably the most famous twentieth-century author who lived, wrote, fished, and drank in Key West. The city sponsors a yearly Hemingway look-alike contest. A Johnny Cash CD played "Ring of Fire" to an empty dance floor.

"What's your handle?" She asked me.

"I go by Stanley Randolf, just call me Stan."

"Welcome to Sloppy Joe's, Stan," She said, then skated over to serve a couple that had just walked in.

Outside on Duval Street, people were scarce—for now anyway. Around 6:00 p.m. that would change. Spring-breakers would be eating and bar-hopping until the bars closed at 4:00 a.m. Country and western music ruled, but jazz and hard rock would contend. I'd had a busy work week in south Florida and was happy to be in Key West.

"This ain't the original Sloppy Joe's," an old guy blurted from across the bar, looking at me. He wore a captain's hat, and pants held up by suspenders made out of alligator skin.

"I know that!" I said loudly, "The original's now called Captain Tony's Saloon—just south of here on Greene Street." I didn't know if he was a tourist or a local.

"I know that!" he answered loudly with a crackle in his voice. "Tell me something I don't know."

I was going to let it drop and not respond. I'd run into many crackly-voiced old men in bars before, and attempts to dialogue with them were seldom worth the effort.

"That Hemingway guy was never in here," he proclaimed, pointing to a large, cloth banner hanging above the dance floor with Ernest's face on it—same image as on Sally's shirt. I took a drink of my Cuban concoction, looked at him, and nodded. He and I were the only customers at the bar.

I knew Hemingway had lived in Key West in the 1930s in a house on Whitehead Street—now a museum. He would write for days on end, sequestered in a free- standing carriage house in the backyard. Then, when satisfied with the words he wrote, he'd take a break and get into serious drinking with his fishing buddies at Sloppy Joe's—the original version. He had a special talent for penning adventure, fishing was often a focal point.

"December fifth, nineteen thirty-three!" the old man shouted, almost spilling his beer.

"Okay, what happened on that day?" I responded, overriding Cash's singing.

"Sloppy's opened—the original." The old guy was coughing now.

"If you boys are going to have a shouting contest, I'm not gonna go deaf from it!" Sally refilled the old guy's beer and moved it across the bar in front of the stool next to mine. That got him to slowly wiggle off his stool, shuffle over, and sit down by his repositioned beer.

"Prohibition's last day was on the fourth," Sally whispered.

"I should have figured that out. That's when the Eighteenth Amendment became history and all of Key West got legally drunk the next day—on the bar's first day." The old man forced a smile at my less-than-brilliant deduction.

"The original, original was Sloppy Joe's in Havana!" Sally clarified. "It had its own brand of twelve-year old rum, popular with American tourists during Prohibition. And served hamburger mixed with tomato sauce on an open bun."

"But some joint in New Jersey named the damn thing a sloppy Joe—not the bar in Havana," the old man grumbled.

I was getting educated, but decided to try changing subjects; Prohibition was ancient history. "What do you think of that guy up there?" I pointed to Hemingway's portrait hanging above the dance floor.

"I never read any of his stupid books," he admitted, looking me in the eye.

"Many others did; he won both the Pulitzer and Nobel Prize for "The Old Man and the Sea," I replied, sipping my drink. "Great fishing adventure!"

"B.F.D.," he proclaimed. "Know what that means?"

"Yup! I know what it means." Sally looked at me, trying to hold back a giggle.

"What, then?" the old man asked.

"There are deals and there are deals; a B.F.D. is clearly a BIG deal." That got a laugh and a wink from Sally.

"It's a big world out there, there's many opinions in it." Sally pulled down on her shirt to straighten the wrinkles on Papa's image. "Papa Hemingway turned into a good guy at the end. My uncle fished with him."

"The egotistical bastard killed lots of big fish—big marlin, big tarpon!" The old man kept his critique on Hemingway going as he exercised his legs on an empty bar stool, probably trying to improve their circulation. I actually felt sorry for him and was amused now that he'd mentioned big fish. He was quite a character and I was starting to figure him out. At least his attempt to stir up a conservation didn't include the weather.

"He did in his younger days, he released them when he got older. Big or small, they were just fish. My uncle was with him when he released a huge swordfish off Cuba—over a thousand pounds," Sally wanted to set the record straight. She pulled down on her shirt to straighten Ernest out again—then patted him on the head.

When the stereo began playing Jimmy Buffet's Margaritaville, I ordered the old guy another beer and a refill on my Cuba Libre. Tarpon fishing was in the cards for tomorrow. The guide I'd reserved told me to expect some serious arm exercise. Today was a do-nothing day—rare for me, but an important mental disconnect for the world-traveling biochemist. Work and adventure periodically required some spacing.

The old man lifted his glass and bumped it with mine. I returned the gesture.

"Call me Ishmael!" he held out his hand to shake. I shook hands with him, he had a strong grip. Then I realized we hadn't introduced ourselves; I had to think fast.

"Call me Captain Ahab," I replied. Sally laughed.

"You're too young to be him." He took a drink of his beer.

"And you're too old to be Ishmael," I asserted.

"His name is Ralf . . . he does this to everyone. Many ignore him. We sell copies of Moby Dick in the gift shop—I think it's the only book he ever read." Sally introduced us: "Ralf, meet Mr. Stan." I shook hands with the old guy again.

I had concluded that his harsh, raspy, cantankerous declarations were a ploy to test people before agreeing to a conversation. I'd been qualified! My instincts must have anticipated this. I'd quickly dropped any inclination to dismiss him after our first minutes of yelling to each other.

Ralf was in his mid-eighties and no dummy. He was an army vet from WWII who moved from Michigan to Key West thirty years ago. We chatted about the Keys; he knew them well.

"What do you know about tarpon fishing down here," I asked him.

"A big one almost drowned me in 1984."

"What happened?"

"Fishing line wrapped around my shoulders when I tried to unhook the fish—I got pulled in the water. When I started to go under, I had to bite the line off; broke a tooth." I could see one of his front incisors was missing.

"Glad you survived. I've learned to always wear a small pocketknife on a chain around my neck when deep sea fishing. How big was big?"

"Nine feet, over three-hundred pounds. A long, thick bugger."

"Wow! That would have been a world record! How long did you fight it?"

"It fought like hell for three hours!"

"What kind of bait did it hit?" Now we were having a meaningful discussion! Sally, seeing where the conversation had headed, pretended to cast out a fishing rod and reel, then had to go refill a pitcher of beer for one of the three tables that had customers. The music switched to George Strait's "Amarillo by Morning."

"A live pinfish held up by a small balloon." Ralf had totally surprised me by this time; the irritable old coot had morphed into a fishing professional.

"Did the tarpon mostly stay deep?"

"Yah, it didn't jump like the smaller ones, eighty pounders. Just two big jumps when first hooked." Ralf's face was bright with excitement when explaining this; he looked ten years younger than when I first saw him thirty minutes ago.

"How did your arms feel after three hours?"

"Weak, they couldn't have lasted much longer. You going tarpon fishing?" Ralf asked me.

"Tomorrow. I have a reservation with the Silver King charter out of Key West Bight Marina."

"How many are going?"

"Just me and the guide. His name is Jones, I've been with him before." I could see envy developing in Ralf's face.

"I'm eighty-six years old . . .. I haven't fished in ten years. I have Parkinson's."

I looked at him, his face had morphed back into vacant stillness. My soul voice spoke to me. I paused to listen . . ..

"There will be plenty of room in the twenty-four foot Mako. You're welcome to come along!"

"You serious?"

"I'm serious."

"I can't walk or stand good, but my arms are okay."

"You can sit in a chair and fish."

"Release all of them?"

"Absolutely. We'll release all of them."

"I have some small money in the bank; I can pay."

"You won't need to. The guide has already been paid."

"You really serious? You gonna take this ancient army vet tarpon fishing?"

"Really serious." My soul voice was resonating at high frequency. I felt a tickle of electricity shoot down my spine . . ..

Jones, our guide, was happy to pick Ralf up at his residence; I planned to meet them at the boat slip at 7:00 a.m. It was a fine Key West morning, and I was primed to crank on some big tarpon. I stopped at a Cuban café and picked up subs for lunch—Jones would provide the drinks.

Inviting Ralf along was one of those spontaneous decisions I had a reputation for making. Some were winners; some were losers. Strangely, I felt good about this one.

Jones eased the Mako out of the dock area, then increased the RPMs taking us straight out into Key West Bight. I could see patches of Sargasso weed floating in the water, some thicker than others.

"Wanna stop and cast for chicken dolphins?" Jones asked. I was standing next to him holding onto the amidships steering console. The twin two-hundred horsepower Mercs propelled the twenty-four foot Mako effortlessly.

"Maybe later. Let's go right after those big tarpon!" Small mahi mahi, three-to-six- pounders, were called chicken dolphins—great eating size. They hung out under patches of Sargasso, feeding on small fish and crustaceans that fed on zooplankton inhabiting the seaweed.

"Sounds good to me." He turned northeast toward Dredgers Key, keeping the long Fleming Key behind us. I could see the Navy Seals training facility in the distance on Dredgers.

Ralf was sitting up front and truly looked ten years younger. He was having what appeared to be a happy talk with himself.

"How did you do yesterday?" I asked Jones. I knew he had taken a young couple out.

"They got two tarpon, about eighty pounds each. Released both. That was enough for them—their arms were sore! We were chumming, using sixty-pound monofilament on spinning reels. Could have easily caught several more, but spent the afternoon trolling the inside reefs for barracuda and mutton snapper."

"See any serious Megalops?"

"There were two over one hundred fifty pounds! They gobbled up the mullet chum as fast as I could cut it. Those big tarpon were swirling in circles just ten feet from the boat. The young man couldn't seem to flip a baited hook into the floating chum. As you know, accuracy doesn't matter on some days, but when the fish are spooky it does."

"Those two big females gobbled most of the chum. I told the guy to flip the baited hook right into the center of it whenever I threw a handful in the water. He couldn't seem to do it accurately. Those smart mamas could identify the hooked piece and avoid it if wasn't disguised well enough. The gal actually hooked both fish, she let hubby fight the second one."

"How deep were you?"

"Twenty feet; right where we're headed now, five hundred yards out from the Navy Seal's training docks."

I put on my psychedelic silk neck protector and pulled it up, covering the bottom of my face as well. "Now I look like you," I told Jones. He smiled, tugging on his own. "They're better than one-hundred SPF lotion!"

I'm primed, ready for action!" I was excited to finally be fishing.

"You're gonna have it!" He assured me. Chum fishing for tarpon in Key West Bight (main harbor area) in March or April was a ninety-nine percent sure bet. How many you'd catch depended on your will and arm strength. The boat in the slip next to Jones's had caught seven the previous day.

"Can Ralf stand?" Jones asked me.

"Not for too long—maybe fifteen minutes."

"We'll get him comfortable with a pinfish on a balloon. He can fish that sitting down. Let's see how you do chumming for those big gals this morning. This afternoon I have a secret weapon for you to try—a new lure you can cast off the bow . . .."

The GPS beeped when we reached the specified waypoint, and Jones cut the motors.

"Ralf, are you busy up front, there?" I kidded him. He was standing up studying pinfish in the live bait well.

"Yeah! Busier than a one-legged fella in an-ass kicking contest. I'm trying to net the biggest pinfish, and he ain't cooperating." Both Jones and I laughed at his analogy.

"I'll give you a hand." I grabbed a second net and went forward to help him.

Of the half dozen pinfish, one big five-incher stood out. It was silver with dark bars crossing horizontal stripes and had the proportions of a large Minnesota pan fish. It was going ninety in the bait well. Two nets made the difference.

Ralf, with pinfish in net, followed me back to Jones, who had the balloon rig ready to go.

"This pinfish is bigger than the one I used twenty years ago to hook that three-hundred pounder—the one that tried to drown me!" Ralf exclaimed.

"They love live pinfish; it's a special treat for them." Jones skillfully inserted an 8/0 circle hook under the backbone of the pinfish. Then, looking at Ralf, Jones handed him the spinning rig—he didn't have to say anything more. Ralf checked the drag setting on the reel, then lofted the balloon with the pinfish into the water—twenty feet out from the back of the boat. Then sat down on a deck chair and opened the bale on the reel. Jones and I applauded.

Fishing tarpon with balloons was not anything knew. It improved the odds of catching big, easily-spooked females that refused to come close to the boat. Ones that had learned how to avoid tarpon fishermen. Jones used medium-sized Shimano spinning rods and reels. A football-sized toy balloon was rubber-banded to the line four feet above the bait. When cast into the water, the balloon kept the baitfish high as it drifted away from the boat. That's it. The secret was in the type of hook used. Since the advent of the circle hook, seventy-five percent fewer tarpon were lost during the fight. A caught fish was one that was brought boat side in the water, then released.

"I have arrived at heaven's door," Ralf declared, sitting back holding his fishing rod and wearing his captain's hat above an oversized pair of sunglasses. Papa Hemingway would have been proud of him, but I knew not to mention it.

Once Ralf was comfortable up front with his pinfish swimming along four feet below the balloon, I was ready to fish off the back of the boat. My spinning rod was rigged the same as Ralf's except with a smaller 6/0 circle hook and no balloon. The hook was tied directly to the sixty-pound test line—no leader.

I baited the hook with a two-inch piece of the mullet that Jones had been cutting up for chum.

"Ready?" Jones asked me.

"Go!" I answered. Jones threw a handful of cut mullet into the water eight feet out from the stern. I flipped my line's hooked piece right into the middle of it, leaving the reel's bale open. We were marking tarpon ten feet deep on the locator.

Line came off the reel slowly as the chum separated and drifted out; my piece kept up with the drift. When the chum dispersed about twenty feet away, we could see several small tarpon surface and consume it.

After the third repeat of the process, with three handfuls of mullet out and significant mullet juice now in the water, two large glistening, silver-green tarpon began engulfing the chum. My hooked piece was in the middle of the action. We could clearly see the two fish in the water. Then a third, much bigger tarpon, suddenly cut them off. Its bucket mouth inhaled half of the remaining chum, including my piece.

"*Wop, wop, wop*." Line came off the reel fast.

"Don't set yet," Jones ordered. "One thousand one, two thousand two, three, four—now!" He yelled.

"I slammed the bale shut and pulled back with the rod, reeling simultaneously to set the hook. Fish on!"

As the fish raced out, line buzzed off the reel, sounding like a score of angry African wasps. Then, about thirty yards away, it skyrocketed out of the water.

"Nice one," Jones declared. "Over six feet—hundred-fifty pound caliber."

The fish bolted to the north while line continued to buzz off the reel. There was nothing more to do in these first minutes but hold the rod high and let the fish feel the power of the fiberglass rod and the reel's drag mechanism.

She greyhounded high in the water for fifteen minutes, running back and forth, north to south, staying perpendicular to the boat fifty yards out. Most of the reeling I did wasn't bringing line in—just reeling against the drag. I looked at Jones.

"Don't increase the drag," He'd read my mind. "It's set at fifteen-pounds. That's equivalent to the fish pulling a fifteen-pound sack of potatoes through the water. Tight enough for spinning tackle."

After another fifteen minutes I was able to do some pumping with the rod and gain line, bringing the fish within twenty yards of the boat. She had been staying high, just under the surface, reflecting daggers of sunlight at us. When she broke water ten yards out, bright sun flashed off the silver on her body like it would a mirror. Seeing the boat, she dove deep.

"It's twenty-two feet deep here. She'll try to follow the bottom into deeper water. This is a smart fish." Jones could tell. "You'll need to flex those arm muscles."

I was standing at the stern with the rod buckled in half; the fish was challenging the drag, taking out line in spurts. This was some serious isometric exercise for my arms. Down with the rod, reel in two feet of line, up with the rod, fish takes back two feet of line. Repeat. Another fifteen minutes passed. We had drifted farther out in the Bight, heading toward the northern tip of Fleming key.

"Tough work." Ralf had reeled in his balloon rig and it was in the water by the bow. He was standing behind me. "Want some help?"

"No, thanks." I was too stretched out to laugh. It felt like I was trying to pull up a 150-pound boat anchor, one that was moving away from me.

"Don't sell your soul; it's only a fish." He patted me on the back. "Bring it to the bow when it comes in. My pinfish is in the water there; let's see what she can do hooked on two rods!"

I grimaced, but inside I was laughing—I knew he was kidding. Jones strapped a rod cup belt on me so I could put more back into the fight. Another fifteen minutes that felt like an hour went by. I was starting to gain on the fish . . ..

Right as I brought her within a few feet of the boat, the big tarpon broke the line! It had taken ninety-five minutes to get her in. Technically, it was not a "caught fish" by I.G.F.A. (International Game Fish Association) rules, which require it be brought to the boat and unhooked. But I was happy; it was my largest tarpon—six feet, eight inches long and fat through to the tail—165-pounds, according to Jones. She was free now, back swimming in Key West Bight and surely not missing that strange fifteen-pound force that had constrained her. And, being further educated from the experience, she would be more cautious of tarpon fishermen in the future.

We took a break for lunch and ate the subs. There is no better boat lunch in Key West than a spicy Cuban sub washed down with a cold Coors. Three yellow-headed, brown pelicans flew by hoping we had some nautical nutrition to share. Jones threw several mullet skeletons in the water, which they promptly dive-bombed.

White pelicans are a different species, larger than the browns. They scoop-feed with their bills while in the water, but do not dive-bomb for food.

"I saw a brown one swallow a three-foot cobia once," Ralf remarked.

"A dead one?" I asked him.

"No, the pelican was alive."

"Five stars for that reply!" I laughed.

"Your friend Ernest claimed dead pelicans looked silly," Ralf recalled.

"I thought you never read any of his books?"

"Don't believe everything you thought," Ralf chuckled.

We went back to where we'd started in the morning, and I caught two tarpon casting Jones' secret lure. Both were brought to the boat and officially released. They were seventy and ninety pounds. The secret lure was a white, foot-long rubber worm—fat-headed and thin-tailed. It had been presoaked in a jar of sardine juice. I fished it with sudden jerks and stops, followed by several seconds of fast reeling, then more jerks and stops. This got the worm's tail to almost touch its head, resembling an ocean creature yet to be discovered. Rigged Texas-style with a 9/0 circular hook, it was durable and easy to fish. Those tarpon slammed it without hesitation!

During the time I was catching those tarpon, Ralf, with unwavering confidence in his pinfish, held his fishing rod under his arm and over his legs while sitting back in the deck chair with his thumb on the reel's spool. He had replaced his captains hat with a wide, straw sombrero.

"Looking pretty comfortable, Ralf," Jones proclaimed.

"I'm at heaven's door—waiting to get in."

"How's that big pinfish doing?" I asked him.

"Alive and kicking, like me. I nicknamed him Moby."

"There's still five fresh ones in the bait well," Jones reminded him. There was a moderate chop in the water now, from a west breeze.

"Moby's still full of energy, he can pull that balloon against the wind."

"Okay, but I can easily set you up chum fishing," Jones suggested. "It might improve your odds today . . .."

"No thanks, I'll stick with Moby."

I tried to help Ralf out by casting the worm lure to the right and left of his drifting balloon. It was a musky-fishing strategy I had used in Canada many times. The objective was to get the attention of a big fish with an artificial lure while reeling it close to a live baitfish on another rod. Quite often a musky would follow the lure, curious, but wary of it, but not wary of the live baitfish it had been alerted to. This might have been the first time the technique was tried for tarpon . . ..

With Ralf hiding behind his magnum sunglasses and straw hat, I couldn't tell if he was dozing or not. I kept casting.

"I'm not sure I approve of what you're doing!" he suddenly blurted. I explained the technique to him.

"Just don't get that crazy worm too close to Moby!"

"I won't, you have my word as an Eagle Scout." Just then, I noticed a disturbance behind the worm lure. Ralf sat up straight—he saw it, too.

"*Wop, whop, whop,*" line rapidly jumped off his reel. Look at this!" He exclaimed, pointing to the reel. With the bale open, the line came off in wops without resistance.

"Don't set the hook yet," Jones yelled, running up from the back of the boat. He could see the balloon rapidly moving off the starboard bow. He counted out eight seconds, then hollered, "Set now!" Ralf slammed the reel's bale shut and pulled back on the rod, reeling fast. The tarpon skyrocketed out of the water, flipping 360-degrees before plummeting back, then raced on the surface, throwing spray while bashing through the choppy water.

"Heaven's door just opened for me!" Ralf proclaimed loudly.

"It's six feet and thick, over a hundred-pounds," Jones affirmed. Ralf knew to not overreact; he kept the line tight and let the reel's drag buzz line out.

"If the fish turns and comes at the boat, reel like hell. Don't give it any slack!" Jones instructed.

"I know that!" Ralf affirmed.

"If it jumps again, extend the rod out, and stop reeling for a few seconds," Jones added. "This is the only time we give a tarpon any slack—to reduce line strain.

"I know that, too," Ralf's voice crackled. The fight was on . . ..

I was truly amazed by the old man. He would stand for ten minutes, leaning on the gunnel, then sit down holding the rod high, then stand back up. All the time keeping pressure on the fish. Several times he brought it in close only to have it turn and go back out.

"Want a drink of beer?" I asked him.

"Thought you'd never ask." I held the can of Coors for him. "Don't be touching my line—that would disqualify the catch."

I grinned. "I won't, I assure you." The old man smiled.

"I can always convert back to cantankerous if you do something stupid!"

"Don't want you doing that. You've already qualified me once!"

"That was a tough run for me with you and the girl defending old Ernest yesterday." He grunted. "More beer." I gave him another drink.

The fight became a standoff between fish and fisherman. The tarpon would take line out and turn, Ralf would reel fast and tighten the line. This went on for thirty minutes.

The fish then went deep, pulling straight down. The old man stood up to get better leverage, his rod bent in a 180-degree arc.

"I wish I had the muscles I had twenty years ago; I'd have this tarpon in by now." The fish did not want to come up. Ralf kept pressure on it, but line continually pulled out against the drag. I tried to imagine doing this at age eighty-six. I could feel his arms straining . . ..

The old man finally caught the fish. Jones, after three tries, was able to grab it and remove the hook. It was six-feet, two inches long—a hundred ten pounder. Ralf was elated; so was I. It had been a four-fish day—not too bad. The old vet's performance was amazing.

Sally called me a month after I'd returned to Minnesota. Ralf had passed away in his sleep. Before he died, he told her to thank me again—and that he would always remember me. I felt that warm, spine-tingling sensation my soul voice uses to comfort me at such times. I paused to reflect on our day together—I could imagine him attempting to qualify a bunch of angels before agreeing to have a serious conversation with them.

Postscript

For those readers who don't fish but still want to have a great tarpon experience, there is a way to do it without driving all the way to Key West: Stop at Robbie's Marina in Islamorada just beyond Key Largo on U.S. Highway One. It's where in you can hand feed dozens of big tarpon off their boat dock—and have lunch too, if you're hungry.

Here's Robbie's story: It all started in 1976 when Robbie and his wife Mona discovered a wounded tarpon in the bay near their dock that had one of its jaws torn open. They brought it in and began rehabilitating the big fish in an oxygenated tank used for shrimp. It was obvious the jaw needed special attention, so they called an old veterinarian friend who came out and sutured it using twine and a mattress needle. After several days of hand feeding and six months of TLC, they released "Scarface" back into the gulf.

Not long after this, Scarface somehow convinced his tarpon buddies to visit Robbie's dock with him in hopes of finding the easy meals he remembered.

Now, decades later, over a hundred tarpon visit Robbie's dock daily. Reflecting the sunshine in flares of silver and green, they inhale the sardines that tourists throw to them. It's a unique place—watch for the "Tarpon Feeding" signs.

Newspaper clipping of Tony Rizzo and author with a 32-pound musky caught in Northern Wisconsin, July 1986
Tony Rizzo casting for Muskies in northern Wisconsin, October, 1992

Chapter Fourteen

Remembering Tony Rizzo

Famous Guide with a Mission

I fished with Tony Rizzo for over twenty years, from 1979 into the Twenty-first Century, strictly for musky in northern Wisconsin. He and his wife Lee owned and operated the Silver Musky Resort and guide service on Star Lake in Vilas County. I became one of his more attentive clients early on, wanting to learn everything he could teach me about musky fishing. I readily recall the treble pitch of his voice when he instructed me on what to do when I had a big fish on. We quickly became good friends.

Tony exuded passion and was clearly dedicated to the sport of musky fishing; it wasn't uncommon for him to fish ten to twelve hours on a guide date, client willing. At five foot six, he was just under five inches taller than the world record musky. But don't confuse length with strength; Tony Rizzo, with his Sicilian roots, was one strong *pescatore*.

One thing that set him apart from many other guides was his straight talk and honesty. Opinionated, yes, but he didn't distort or exaggerate; if he got skunked several days straight fishing muskies, he'd admit it. When Tony caught a big musky, he would never think of pouring sand into its belly to increase weight or taking pictures wearing several different shirts to suggest multiple catches. Such foolery plagued musky fishing in the early days. Some taxidermists back then even inserted spacers in their musky mounts to add extra length. In those days a trophy musky (forty-four inches or more, according to Tony) meant $25,000 in revenue to Vilas County, Wisconsin; the average spent by a tourist angler before catching one. This included lodging, meals, gasoline, tackle, guide fees, etc. So, be advised: musky fishing is a serious business!

Tony Rizzo kept impeccable records of water temperature, wind direction, musky follows, muskies boated, muskies missed and lost, and their various sizes. In his *Secrets of a Musky Guide II,* (F.A. Weber and Sons, 1990) he presented the following statistics representing his first twenty years as a professional musky guide:

- Time on the water: 20,000 hours.
- Undersize muskies boated (less than thirty-two inches): 1,589.
- Muskies thirty-two inches or larger boated: 1,514
- Muskies seen but not hooked (follows and misses): 1,500.
- Muskies lost (hooked but not boated): 495.
- Muskies forty-four inches or larger: 100.

Tony considered a forty-four inch musky or larger a trophy, commonly weighing around twenty-five pounds. His figures reveal that two hundred hours on the water was required to catch one. It's noteworthy that the Wisconsin Department of Natural Resources reported in 1990 that only 1.9 percent of the musky population reached that size or larger. Big muskies are few and far between, a significant challenge to find and catch, even when you know what you're doing. They've been referred to as the fish of 10,000 casts. I used to joke with Tony that it only took 9,500 when you knew what you were doing.

My First Musky—July 1975, Before Meeting Tony

They call it "musky fever." It's a malady that can express itself with a variety of psychosomatic symptoms, ranging from a mild elevation of body temperature to a serious inability to think about anything but musky fishing. For those infected there is no cure. The only treatment that temporarily offsets symptoms is musky fishing.

Musky, *Esox masquinongy,* is the largest fish in the pike family and is native to northern and eastern North America. The current all-tackle world record is sixty-seven pounds eight ounces with a length of 61¼ inches. It's every musky fisherman's dream to break that record.

I'd been fishing muskies since 1975, four years before meeting Tony Rizzo. My first musky was caught on Lake of the Falls in Mercer, Wisconsin in July of that year, casting a wooden lure—a red and white Lazy Ike. The fish weighed eight pounds and was thirty-two inches long—a legal keeper at the time. We ate it.

Driving home to Milwaukee back then, I kept thinking about that musky. I had previously caught various species of large ocean fish, but the sensation wasn't the same. That eight-pound musky followed the Lazy Ike right to the boat and swam around it a couple times, studying it. When reeled fast, the lure has an exaggerated head-down wobble. I had read enough about musky fishing to know to keep the lure moving as it approached the boat on a retrieve. Muskies are the top predator in northern lakes and rivers—fish-smart, wary, and not easily fooled. I recall making a large figure eight with the lure as the fish neared the boat. Without warning, it jumped and exploded on the lure, hooking itself, then jumped again, completely breaking water before landing in the boat. I never had to use the net. I was feeling amazed and feverish as I drove home thinking about musky fishing—it appeared I was pretty good at it . . ..

My Second Musky—July 1979

Between July 1975 and July 1979 I caught walleye and bass in northern Wisconsin, but not another musky—a four-year dry spell. By this time, that first musky felt more like a lesson in pure luck rather than any proof of knowledge or skill on my part. In July of 1979 the owner of a bait shop in Boulder Junction (B. Jct.), Wisconsin advised me to hire a guide. He recommended Tony Rizzo, who I'd heard good things about. I called Tony from the bait shop; he answered on the first ring. It was 6:30 p.m. and he had just returned from guiding. I introduced myself and told him I was up north and needed a good lesson in musky fishing.

"You're in luck, my friend—my party for tomorrow had to cancel: sickness in the family. So I'm available."

"Wow, great!" I was pleasantly surprised. I hadn't expected this to happen so fast. We discussed the details of what I needed to bring in the morning, and that I should meet him at his Silver Musky Resort on Star Lake at 8:00 a.m.

"You're very lucky," Walter, the bait shop owner, told me. "Tony Rizzo is the best musky guide up here, and he's booked solid this time of year." I looked around as Walter kept talking. He was happy to show me a wall full of musky lures. I knew that Boulder Junction was trademarked The Musky Capital of the World, and this bait shop was right in the middle of it.

"You'll need more than just a Lazy Ike," Walter advised. I was happy to let him suggest a few options. Of course, it didn't take much convincing for a few recommendations to turn into seven new lures sporting a full spectrum of sizes, shapes, and colors: An orange bucktail, a black Suick, a silver Rapala, a yellow Cisco Kid, a white Globe, a brown Eddie, and a red and white Daredevil. It was clear that getting into musky fishing required making an investment and learning a new vocabulary. I paid a smiling Walter, then drove three blocks to the Guide's Inn and had dinner: lamb chops, scalloped potatoes, and Point Special beer. A large mounted Muskellunge stared me down as I ate.

In spite of my excitement, I slept like a rock at a local motel that night. In the morning I stopped at the old grocery store in B. Jct. and bought four ham sandwiches, two apples, two candy bars, and two Cokes. Tony had told me to bring lunch for both of us, we would be eating on the water. Brown bag lunch in the cooler, I headed east on Vilas County Highway K to Star Lake—about a forty-minute drive.

Tony was filling his boat's gas tanks when I arrived. He was dressed for the occasion: white tee shirt, beige shorts with a smudge of grease, and a Tuffy Boat hat.

"Hey, Mr. Stanley. Pleased to meet you." Tony shook my hand with his left hand while continuing to operate the gas pump right-handed. I returned the greeting.

"Put your tackle in the boat. We'll get to know each other on the way to Plum Lake." I appreciated Tony from the start. He was Mr. Musky, no B.S. He got right into a dialogue on how to catch *Muskel-*

lunge.

The seventeen-foot Tuffy followed two feet behind us as we took the dirt road that ran past his resort to the lake. "There's action lakes and trophy lakes up here," he said, obscuring our rearview with a cloud of North Wood's road dust. "The action lakes produce more muskies, but they're smaller than those from trophy lakes. Today we'll start with an in-between lake that has both action and big fish."

"Sounds like a plan," I remarked. It was a splendid blue-sky morning in the North Country. Tony was short, requiring that he sit forward on the front seat with his eyes glued straight ahead as we bounced along on the narrow road.

"Had a six hundred pound black bear run out last week on this road; almost hit it. Did you strap your two rods down?" He asked without looking at me.

"Yup, with two bungee cords."

"Good man. And lunch?"

"It's in a small cooler, also strapped."

"Excellent. I don't waste time going to a tavern for lunch, like some guides. It gives us more time on the water. Plum Lake is a thousand acres—maximum depth, sixty feet with an average of twenty. Sand and gravel bottom, not much muck. It's a great summer musky-lake." Tony checked the time. "There's a four-footer that's active now on a mid-lake bar north of the boat ramp." It was 8:20 and I could tell he was antsy to get the boat into the water.

After he backed the Tuffy in, I held it by the dock while he parked the car.

Before heading out he wanted to check my fishing equipment. He checked the line guides on each rod with a Q-tip. No strands of cotton pulled loose when he passed the Q-tip through them, so the line guides were good—no burrs. He made me retie double-improved clinch knots on the leaders on both rods. I opened my tackle box and showed him the lure investment. He nodded, seeming to approve. Looking around I could see an assortment of well-used lures stuck to the boat's carpeting—lots of them.

"Careful where you step!" He exclaimed in his signature high-pitched voice as he handed me one of his Rizzo Tails—a wire with chewed-up black feathers wrapped around it and a silver French spinner blade at the top. It was not fat with deer hair like the bucktail I bought from Walter, but thin, and frankly quite ugly.

"Start with this black Rizzo Tail. We'll get to your fancy lures later. Reel as soon as the bait hits the water. Reel fast. We'll be fishing over deep weeds. If you feel anything, set the hook hard and keep reeling."

We got to the mid-lake bar by 8:40. He quickly put the electric trolling motor down and started running it from the back of the boat. I had already snapped on the Rizzo Tail and started casting from the front of the boat, making sure to follow his instructions. I was ready to beat my average of one small musky every four years.

The ugly lure came alive in the water. Its stringy black feathers were from a marabou stork and were naturally waterproof—they danced on the wire, breathing like they were alive. I immediately changed my mind about the thin, fast-moving lure. It looked more like a large leech rather than some rodent—what fatter bucktails with deer hair were claimed to resemble. Muskies ate a mixed bag of North Wood's critters.

On about my ninth cast, I got a hard hit. We were well away from shore, over the deep weeds. I set the hook and knew it was a big fish. Those first few seconds when nothing moves, not you or the fish, then the strong pull of its powerful run—is unmistakable. A big Florida snook on spinning tackle was my best frame of reference at the time.

"Keep reeling fast!" Tony shouted. I kept reeling and could feel the weight of the fish swimming deep and pulling hard.

"It's the four-footer!" Tony said excitedly. I kept reeling; my mind totally focused on the big musky.

"She's under the boat!" I cried out, leaning over the side to get more leverage just as she broke water behind me, on the opposite side of the boat. I thrust my rod into the water up to the reel to keep the line from snagging on the bottom of the boat—instinctively I knew that much—then repositioned myself to be on the same side as the fish. The musky headed back out, away from the boat.

"Holy Jesus. A large musky on my ninth cast!" I reeled fast enough to keep the pressure on her but not overpower the Garcia 5500 reel. "She's taking drag now," I blurted as line rushed out.

"Reel faster, that's a good reel. It won't break. You're doing great." I could see Tony was moving the boat with his left foot on the electric motor. Taking us out into deeper water.

The fish stayed deep about halfway back to the boat, then surfaced, splashing and slashing its head trying to throw the lure. "You've got a good hook," Tony shouted. The front treble hook on the lure was embedded in the right corner of the musky's mouth. She kept opening and closing her mouth, trying to get rid of it. I could see how well the hook was holding in the tough, leather-like tissue in the corner of her mouth.

"She's down deep again . . . pulling hard!" I exclaimed.

"That's because she's swimming away now. When she stops pulling, she'll be swimming back at you. Don't give her any slack; reel extra fast when she starts coming back . . .. This is a thirty-pound class fish!"

She kept pulling for another five minutes. I was standing on the bow, holding the rod high. Tony was ready with his oversized net. What a rush this was—both physical and mental.

"Bring her in head first." Tony was standing to my right. "It's over twenty-five-feet deep here. I got us off the bar and a patch of thick coontail (weeds). Bring her in."

He held the bag of the net below the surface and swept her head into it as I nudged the fish forward. Then the entire fish was in the net, a huge, throbbing U-shaped muscle.

The fish was a fat, forty-six inches and weighed twenty-eight pounds. Not bad for a second musky that ended a four year dry spell. I stood up holding the fish, Tony snapped a picture.

Tony was all smiles as we sat in the boat staring at the fish. "Do you know the name of that mid-lake bar where you caught this fish?

"No, tell me."

"It's called Stanley's Bar."

"Really, since when?" I suspected he was putting me on.

"I'm serious; since now!" Tony asserted. "You're the first client of mine to catch a trophy musky on it. That's how I name reefs, bars, and bays up here."

"I'm honored," I said, humbled.

"Just keep catching those big fish. By the time the Good Lord calls me to guide him on Heaven's Lakes, I intend to have used your name on many of these Vilas County lakes.

"Let's get back at it." Tony smiled big.

Around 11:00 a.m., I missed a thirty-six inch musky in thick weeds, just before Tony caught one about the same size. In addition we had several big follows. At noon we ate the ham sandwiches and drank the Cokes in the boat. The morning had been all I'd hoped for and more.

"We don't have to go far to get to the best taxidermist in the North Country," Tony declared as he took a bite of his sandwich. "Neil Long's studio is just off Highway N—you can almost see it from here." He pointed west to where a bridge crossed the lake.

In 2019, the twenty-eight pound musky still hangs on the wall of our family room, which is now in Minnesota.

On the way back to Tony's Resort I asked him how Plum Lake got its name: "From some guy named Plum? It makes me think of Professor Plum in the game *Clue*."

Tony laughed. "It's named after Madam Plum who befriended the CCC lumbering crews during the depression."

"Befriended? That's a good way to put it." I grinned.

"Prostitution was big business back then. Madam Plum was a business lady in control of much of it. The lumber crews would party on Saturday nights, then go musky fishing on Sundays. Many of the lakes up here are named after their favorite girls. Laura, Eloise, and Dorothy Dunn are a few examples."

"Dangerous combination: prostitution and musky fishing. Either one could do you in under the wrong circumstances," I added.

"I'll stick with musky fishing," Tony assured me.

"Ditto," I agreed. "I have an old photograph at home that my grandfather gave me; it shows two lumberjacks in northern Wisconsin holding a rope strung with at least two dozen muskies. It was taken in 1934."

"Those were the good old days," Tony commented. "How many trophy fish were hanging on the rope—any larger than forty-four inches?"

I had to think for a moment. "Now that you mention it, maybe one. Most were two to three feet long."

"See, even in those good old days trophy fish were rare," Tony emphasized.

"Tony, you're the man!" I gave him a bear hug. "Can't wait to do this again."

White Sand Lake—August 1982

I had many guide dates with Tony during the 1980s. His favorite time to musky fish, all things considered, was the first warming period after the first hard frost—usually occurring in September, some years in August. This signaled the big mamas—"super-muskies," Tony called them—to eat like there was no tomorrow as they stored up fat for winter. Tony defined a super-musky as one approaching, or exceeding, fifty pounds. I almost caught one with him in August 1982 on White Sand Lake, just east of B. Jct.

"Tony, look at this fish!" I declared, not too loudly; I didn't want to spook it. My fishing rod was bent in half, and a giant fish's mouth was clamped shut on the yellow Cisco Kid I'd been casting. The big musky was hardly moving, just fanning its pectoral fins, suspending in the water off the bow, biting on the lure, staring at me.

"Stanley, she's not hooked!" Tony whispered in a soprano pitch.

"I know it!" We could see the lure's two treble hooks hanging free, as the fish gripped the body of the lure between the hooks. She felt no pain.

"Keep pressure on her; she might get hooked when she lets go of the bait. With no hooks stinging her, she's relaxed about it all. Curious."

My heart was pounding. I kept the rod bent; no need to reel—the monofilament line was as taut as strings on a Louisiana banjo. When I pulled her slightly closer a few inches, she shook her head and backed up, recovering those inches.

"She's huge Tony, this is unbelievable! The monster kept staring at me.

"She's a record-class fish—maybe sixty pounds. Keep moderate pressure on her; don't pull the lure out of her mouth. What a super-fish!" The tug-of-war went on for several minutes, seeming more like an hour.

I could see the excitement in Tony. I was excited too and astonished as well. Standing on the bow, I held the rod up at a forty-five degree angle. I was transfixed by her eyes, the size of half-dollar coins, and her crimson-red tail and ventral fins.

"She's a real honey! Maybe I can tail-net her." Tony grabbed his large net and stood behind me. The giant musky shook her head hard when she saw the net, but still held the Cisco Kid tight with the hooks hanging loose.

Seconds later, the huge fish opened and closed her mouth, and then violently shook her head. The lure's hooks still didn't set. *How is that possible?* I thought, keeping a tight grip on the rod.

"She wants to kill it, then eat it, like she would a perch or walleye. Try to pull her closer." Tony was kneeling, holding the bag of the net in the water. I gave the musky a tug toward the net. She compensated with an equal tug, backing up again. The massive fish was not swimming or fighting—just suspending perpendicular to the boat, her eyes fixed on both of us.

"It's a musky's version of the standoff at the O.K. Corral," Tony whispered. His hands shook as he held the net. Tail netting was the only option with the lure's hooks swinging loose from her mouth.

I kept the rod high and could feel the throbbing tension the fish produced. "Have you ever seen this before?"

"No, not with a fish this size. This is a first." A shiver went down my spine.

Then, suddenly, she jolted, opening her mouth extra wide. The lure instantly released and propelled back from the suppressed tension—hitting me in the chest. Both hooks pierced my leather vest.

"Oh, no!" Tony shouted. The giant musky's eyes remained fixed on us as it sank. I watched her disappear in White Sand's seventy feet of crystal clear water. A second shiver went down my spine. The fish confirmed Tony's theory: Big muskies suspend over deep water. Casting out here in broad daylight in the middle of a seven hundred fifty acre lake, away from all subsurface structure, is not the way most musky fishermen would spend a summer afternoon.

Darkness approached as we commiserated together. We agonized over what we could have done differently. Maybe another lure would have hooked the fish. Maybe I should have pulled on the rod sideways instead of straight up. Maybe the net spooked the fish. We debated all the maybes.

"This is likely the largest musky I've seen in fifteen years of guiding," Tony groaned. "I say 'likely' because I've lost big ones fishing live suckers that never came up. I'd trade all the big muskies I've caught for all those I've lost—without giving it a second thought." Another shiver went down my spine.

"We proved your theory! Big muskies suspend over deep water away from bars, weeds, or other structure."

"I wouldn't bring just any of my clients out here or out in Trout Lake's hundred feet of water. Most want action. I typically get skunked casting deep water. Experimenting! But knowing you're a research scientist who likes to experiment, I didn't hesitate to bring you. And we just had an unbelievable musky experience!"

As we discussed our evening strategy, the two of us munched on Snickers bars while the sun dropped behind the line of white pine trees hugging the shore. We could see walleye fisherman going out to a shallow sand bar with a bucket of minnows hanging over their boat. I could hear two of them talking and pointing at us. We were still out in the middle of the lake, their voices traveled in the damp air.

"Could you hear what they were saying?" Tony asked me.

"The one old guy told the other old guy: 'Those two chaps are lost, they need a guide.'"

Inventing a New Diving Lure—1983

After the episode on White Sand Lake, I started shopping around for different types of diving lures. I found a jointed, eight-inch lure with a stainless-steel lip, just like the one on a Cisco Kid. The lure was made of cedar wood and called a Master James; it had great action in the water. When reeled in, its two treble hooks would be easier to set than those on the Cisco Kid. I was excited for my next guide date with Tony, ready to surprise him with the new lure. The two hundred seventy mile drive from my home in Milwaukee to Tony's

Silver Musky Resort zipped by quickly as my mind, occupied with thoughts of musky strategies, disregarded time.

I was standing with dirty tennis shoes in the back doorway to Tony's kitchen.

"Come on in, Stanley," his wife, Lee, called out. "Want some toast?"

"No thanks, Lee, I had a big breakfast at Sandy's in Boulder. I'm a little early," I said to Tony.

"Early is good. Want some teen-fish action this morning? There's a small lake up by Eagle River that's been hot." Tony was still chewing on his breakfast toast when I arrived at his resort. He could clearly enunciate a lake's musky features with his mouth full of toast.

"Sure. I have a new diving lure to show you." Hearing that, Tony raised his eyebrows. Not many of his clients, particularly the old guys, experimented with lures. They simply used one of Tony's proven weapons that were stuck all over his boat's carpet.

Tony slugged down the rest of his orange juice. "Go put your gear in the boat. I'll be right out."

"Snipe Lake is a small, mesotrophic, two hundred sixteen acre lake—fifteen feet deep at its deepest. It's full of twelve-to-eighteen-pound muskies that seldom get fished. These teen muskies as I call them, were active last week. We'll hit Big Saint Germain this afternoon for the big gals," Tony explained. I knew the largest muskies were females.

"Rain is forecast for the afternoon," I mentioned, as three whitetail deer crossed the road ahead of us.

"Rain or shine it's still musky time," Tony recited. "Got your raingear?"

"Always!"

We were the only fishermen on Snipe Lake at 9:00 a.m. Tony started casting a black Rizzo Tail, and I launched the brown Master James. It was ten feet deep off the dock on the west end of the lake. To the east the sky was clear, but thick cumulus clouds dominated the far west.

Tony put his hand in the water: "Seventy-two degrees." He smiled. "It was seventy-eight last week." Muskies are most active around 68° F. Cold nights had been cooling the lakes. I checked Tony's accuracy with a thermometer once; he could accurately tell the water temperature with his left hand.

We fished our way out to the center of the lake, Tony in back, me in front. I was pleased with the action of the Master James; it ran five feet deep with a classic head-down wobble. Not as tight a wobble as the Lazy Ike but much more than the straight Cisco Kid—and it clicked in the water due to its jointed construction.

I enjoy casting. I was belting out and reeling in two casts per minute, always making some extra commotion, a figure eight or an L or C as the lure approached the boat, ready to set the hook on any musky following. In dark water like on Snipe Lake, the muskies can come out of nowhere at the last second.

"Here's one," I shouted, rearing back to set the hook. "Feels nice, a strong fish." A fat thirty-five inch, fourteen pound musky clamped down on the Master James, splashing wildly. Tony quickly netted it, held the fish up to show me, then released it. It was only 9:30. By 11:00 I had caught four more muskies between thirty-two and thirty-seven inches—totaling five legal-size fish, released them all. Tony caught one thirty-four-incher on a brown Rizzo Tail.

"Let me see that thing." Tony pointed to the Master James. I could see the wheels turning in his head.

"You're the first client to ever catch five legal muskies with me in one day! Every time I turned around you were setting the hook." He kept eyeballing the lure, swinging it and grabbing it with his hands, listening to the two wooden sections clicking. "This gives me something to think about," he mumbled, nodding to himself. I smiled.

Then the rains came. It's amazing how fast two men can put a boat on a trailer in the rain. We ate lunch in his car on the way to Big Saint Germain Lake: tuna salad subs, pickles, and Cokes.

Big Saint Germain, just south of Sayner, Wisconsin on State Highway 155, is a sixteen hundred acre, eutrophic weed bed with a twenty-foot average depth. I planned to cast a red Rizzo Tail in the lake's thick musky cabbage; a diving bait would snag too many weeds. The trick to not catching weeds with a bucktail is to control your cast and reel as soon as the bait hits the water, then hold the rod up and allow the spinning blade to bulge the water on the surface. Last year I missed a four-footer with Tony in Big Saint's thick weeds at a spot behind the big island. I was momentarily fooling with the zipper on my rain gear. Didn't get a good hook set when the fish hit. It was on for ten seconds before leaping straight up and off. Tony triangulated the spot and went back to it the next day with a different client and caught it—on a red Rizzo Tail. I wondered if he remembered this.

"Red Bucktails with brass-colored spinner blades are a guide's rainy day favorite out here," I remarked; I'd been doing some reading.

"I'm not sold that specific colors make a big difference. There's really two color groups: Bright colors like red, orange, and yellow; and dark colors like purple, brown, and black. Dark lures throw a bolder shadow, bright lures have more flash. So it's good to switch groups back and forth on any day. I like white at night because I can see the lure better."

"Do you favor dark over bright lures on dark days, bright on sunny days?"

"Not really. My records don't show a significant difference. Stay with whatever color feels good for you, but do switch groups occasionally."

We kept discussing lures with emphasis on bucktails. Tony was keen on using bucktails (Rizzo Tails) with different types of spinner blades: Fluted versus French versus Colorado, Indiana, willow, etc. Each had its own frequency of vibration which was more important than color.

As we approached the launch, the rain intensity increased along with the wind. "Let's pull over in the parking lot until things calm down somewhat," Tony said. I nodded in agreement.

"You know I like Big Saint just before or after a rain," Tony reminded me. "And don't fool around with your zipper and miss another big gal this time."

"I won't, you can bet on it!" I smiled. He remembered.

I appreciated his critique. This man guided every day during the musky season, June to ice-up in November, and he not only remembered every musky caught but all those lost too!

I knew the rules of the road by now: To put the odds in your favor, you must set the hooks instantly when a musky strikes; two seconds late often spells failure, no fish. After hours of casting, it's easy to get distracted and lose your concentration.

We sat in the car at the boat launch, as cats and dogs rained down. Nobody was out on the big lake. There was always something to talk about as we waited for a pause in the windy wetness.

"Did you know Porter Dean?" I asked Tony.

"Sure did, went out with him back in 1957. He was called 'The Barefoot Guide.' Made his world-famous fried potatoes at shore lunches. Lots of celebrities hired him: Jane Russell, Janet Gaynor, Gypsy Rose Lee, and Ted Williams. He took Eisenhower out three times before he was president. Ike caught a musky on Twin Lake in Phelps and one on Partridge Lake. The secret service got all bit up by bugs when they camped out on Partridge, guarding Ike. They didn't get a photo of either of Ike's fish, so they had to stage one with a fish caught by Porter. Ike stood on the bow of the boat, holding Porter's pole with a fish hooked on it. *Outdoor Life* put the picture on its cover: "Fearless Ike fights a frightful muskellunge."

"What other lakes did Porter fish?"

"All the ones I fish plus a few more. I concentrate on about thirty. There's five hundred and sixty-three named lakes and seven hundred fifty-five unnamed in Vilas County—ninety-four thousand acres of them. Most have muskies.

"Porter Dean reportedly lost the world record on Trout Lake—at the narrows where North and South Trout join. Estimated at eighty pounds, he tried to club it with a Coke bottle."

"That's incredible. It could give a guy the fever!"

"You can bet on it. Let's go. The rain is slowing."

There are dozens of good fishing spots on Big Saint German Lake. We fished between the big island and shore in fifteen feet of water, about fifteen yards out from shore parallel to County Highway C, then in twenty feet of water off Big Saint Germain Drive. We also finished mid-lake weed beds—only using Rizzo Tails. No action.

"The rain put the muskies down today; last week it turned them on," Tony lamented.

There were no takers on Big Saint Germain, just lots of healthy, aerobic casting exercise. We went out on Star Lake by his resort later that night, and I had a serious strike in deep water on the Master James. Didn't see the fish, but it was big: "made a splash like a horse falling off a cliff," to quote Ernest Hemingway.

"I want to see your version of this bait next season!" I declared, holding up the Master James. I prodded Tony as we finished the twelve-hour day on a weed bar a stone's throw from his resort.

"Stan, my friend, you know me well by now. I won't disappoint."

The following July Tony was ready with a modified version of the Master James (which was no longer on the market). His was made from ABS plastic instead of cedar wood, about an inch longer, somewhat thinner, and virtually indestructible compared to the Master James. It also had a more pronounced click in the water with its treble hooks hitting plastic instead of wood. The Rizzo Diver had been invented, and I was excited to try it out on a guide date with Tony.

We met at the boat launch for Lake Laura on County Highway K east of Star Lake. It's a beautiful, under-fished, 628-acre oligotrophic lake with a maximum depth of forty-three feet. There was one other boat on the lake at 8:30 a.m. when we started casting. Tony kept us on the forty-foot depth contour running the electric motor constantly.

We both pitched Rizzo Divers in the middle grounds of Lake Laura. Unlike the Master James, the Rizzo Diver dove between eight to ten feet down on a retrieve—three to five feet deeper than the Master James. Its action—an enticing head-down wobble—was a superb match for the Master James. Casting for three hours without taking a break, we completely circumnavigated the lake and were ready to repeat. No muskies had followed or hit at this point.

"How about a roast beef sandwich with a Coke?" I suggested.

"Great idea," Tony agreed.

We sat down in the boat, and I opened the cooler. "I hope you still like mayo spiked with horseradish and a touch of garlic?"

"You bet I do!"

The sandwiches slid down nicely. Tony was happy that I now had a residence in the North Woods and could make sandwiches at the family's cabin. Any sandwich was better than those sold at gas stations.

"For a biochemist and budding musky-aficionado, you're not a bad sandwich maker."

"Thank you, Mr. Rizzo. These sandwiches were formulated to entice muskies to hit. But that's only if the fisherperson pees in the lake after eating one."

Tony laughed, "It must be the touch of garlic, I know bass like garlic."

Musky fishing sometimes invites temporary lunacy. It's one of the drawbacks. Catching one settles me down. I was ready.

The fish locator was showing bait fish suspended at ten feet. Exactly where our lures had been tracking all morning. When lunch was over, it was back to casting. I noticed the other boat was still on the lake, one man was casting weeds along the shore.

"Here's a big fish," Tony shouted as he reared back to set the hook. I saw the splash behind the boat, like our eighty-pound Labrador would make jumping off a dock. The fish took Tony to his knees, almost pulling him into the water. Then it leaped completely over him, hitting the starboard gunnel and landing back in the water halfway to me.

"How did you ever get it to do that?" I yelled, net in hand.

"The fish did that, not me," Tony hollered back.

The fish was in the water in front of me, going crazy. The back treble hook on the Rizzo Diver was in its lower jaw.

"What an incredible jump," I declared. "I thought you pulled the fish out of the water over yourself—a new guide trick!"

"No. She did it all—this baby can fly. Get the net ready."

I stood next to Tony, holding the net. The fish was a beauty, forty-nine to fifty inches! It was bull-dogging on the surface in front of us. Before I could calibrate the action in my head, the fish bolted into the air and threw the lure—both lure and fish almost landed in the boat.

"Get it now!" Tony yelled. The fish was momentarily stunned.

I got down on my knees, holding the long-handled net by the rim with my left hand and the net's bag away from the fish with my right hand. The fish was rubbing herself on the side of the boat—she teetered on the rim of the net which was partially under her. I needed another hand. Tony was alert; he grabbed the net's handle while I hung onto the rim with both hands, pulling the fish toward me and into the net.

"We got her!" Tony yelled.

"Wahoo!" I shouted, a favorite saltwater fisherman's exclamation.

We both sat in amazement, looking at the trophy in the net. Then, out of nowhere, we heard clapping. Looking up we saw a lone man in a small boat, standing up and clapping his hands. He was the lone fisherman we'd seen all morning, also a musky man who seemed thrilled to have witnessed the catch. Tony and I saluted his appreciation, and waved him over to see it up close. He wanted to know where he could buy a Rizzo Diver.

I got good pictures of Tony holding the fish: fifty inches, thirty-two pounds. You can see the best one of these photos on the front cover of the fifth printing of *Tony's Secrets of a Musky Guide,* (Willow Creek Press, June 1988). Proof that Rizzo Divers catch suspended summer muskies—big ones.

Birthday Party for Tony Rizzo

Marty Smith, one of Tony Rizzo's most senior clients, organized a party for Tony's seventieth birthday in August of 2005. It was at the Riverstone Restaurant & Tavern in Eagle River, Wisconsin, and would be a surprise. We were told to gather quietly in the restaurant's downstairs party room an hour before Tony and his wife would arrive. Marty was expecting about forty of us old regulars to show up, told us to be prepared to tell musky stories.

Two hundred and forty people showed up! Tony was overwhelmed when he saw us all; tears flowed down his Sicilian cheeks. As far as I could tell, he remembered everyone's name, husbands and wives—musky men and musky women! Strangely, it was the first time in twenty-five years that I'd seen Tony dressed in something other than fishing attire—he looked great in a turtleneck and sport coat.

After several servings of special North Country hors d' oeuvres and a few drinks, we settled into telling stories. The best story would win a 250-dollar gift card donated by Eagle Sports in Eagle River, Wisconsin.

I smiled to myself, thinking of the sixty pound-plus, never-hooked, potentially-record, gargantuan musky on White Sand Lake that suspended for a good five minutes on the end of my line before sinking into the depths, unscathed—it had to be the winning story! Particularly when punctuated at the end by the two old walleye fishermen concluding we were lost in the middle of the lake and needed a guide.

The stories flowed and Tony had a ball. Old memories surfaced—some hilariously embarrassing. With over four thousand muskies boated, forty years of guiding under his belt, thirty-five thousand hours on the water, numerous awards and records, and nine books authored, Tony was without a doubt, a legend in his time. He was the man we all loved and respected—a truly special human being.

I was sure my story had good odds of winning; the more stories that were told, the more winks and thumbs-up I got from musky men and women in the crowd who had already heard mine. I was excited.

But I didn't win. The very last story, told by an elderly lady named Lilly, won the prize . . ..

Here's Lilly's story as she told it:

Lilly cleared her throat. "I'm eighty-three. When I was sixty-three, I caught a twenty-five pound musky with Tony Rizzo. I had never fished muskies before and neither had my husband Lewis, who's gone now. We had the musky mounted and it still hangs over the fireplace in our house." She paused and took a drink of beer. You could hear a pin drop.

"It was raining that day. Tony took us to a lake called Buckatabon and gave me one of his fishing poles to use; it had a Rizzo Wiz lure attached to it. He was very patient with me and showed me how to cast before taking us out on the lake.

"After fishing for about an hour, the Rizzo Wiz got stuck on the bottom. I told Tony his lure was stuck, but before he could help me, it started to move. 'Pull back hard and wind fast!' Tony yelled. I did that, and a big musky jumped out of the water, almost landing in our boat.

"'Keep winding the reel,'" Tony instructed. I did, but my hands were cold and wet. When the fish jumped again, it pulled the rod and reel out of my hands, right into the water—it sank. I'd lost Tony's good rod and reel.

"Tony could see what happened. He carefully looked around for several minutes and kept checking a green box on the floor that had flashing lights.

"'Don't worry about it,'" he told me. "'We'll come back later.'"

"So we went away to a different part of the lake and fished walleyes for an hour. Lewis caught two small ones. After that, Tony started the boat's motor and brought us back where the musky pulled the rod and reel out of my hands. Tony kept looking around and watching the lights flash in that green box.

"Then, suddenly, he stopped the motor and handed me another rod and reel with a big red-and-white metal thing on it. My husband called it a Daredevil.

"Tony pointed to a spot in the middle of the lake, 'Cast there,' he said. So I did. I made a good cast. 'Let the lure sink to the bottom, don't wind the reel,' he instructed, still watching that green box.

"'Now, wind!'" He shouted. I wound the reel but it felt stuck—again. Seeing this he grabbed the rod from me, held it deep in the water, and wound the reel slowly. Then, when he seemed to feel something, he grabbed down in the water. I could see he had caught the first rod and reel. He untangled the two, then lifted the first rod high and started winding fast. The rod bent in half.

"He then handed the rod back to me—'Here, Lilly. Finish catching your musky.' The fish was still hooked on the Wiz and it gave me a good fight until Tony could net it.

It's my favorite memory."

Postscript

Tony Rizzo passed at age eighty-three on September 18, 2018 at the Friendly Village Nursing Home in Rhinelander, Wisconsin. His wife Lee had passed five years earlier. Tony's son, Tony Jr., and his wife Katy, own and operate Anthony's Restaurant housed in the Old Milwaukee Road railroad depot in Woodruff, Wisconsin. Recently, in 2019, he and his wife put a huge collection of his father's memorabilia on display at the restaurant: photos, awards, mounted muskies, tackle, lures, magazine articles, books, and more. Even if you're not a musky angler, I assure you, the great Sicilian food is worth a stop. And who knows, you may catch the fever.

The last time I saw Tony was in 2015 at Helen and Rollie's Musky Shop in Minocqua, Wisconsin. He was restocking the store's supply of Rizzo Wiz, Tails, and Divers. After the hugs and back pats, we each had a cup of Joe and talked for an hour. Before leaving I asked him if there was anything he would do differently, if he could do it all over. He put his hand on my shoulder.

"Three things, my friend:

One: I would spend more time fishing in twenty-five feet of water.

Two: I would fish much more at night.

Three: I would buy a lodge in Ontario, Canada on Eagle Lake."

Visit Stanley, see photo gallery & more at:

www.AdventuresofaWorldTravelingScientist.com

More books by Stanley Randolf:

Adventures of a World-Traveling Scientist

More Adventures of a World-Traveling Scientist

Questions for Group Discussion

1. What was your favorite story in this book and why?

2. What did you find the most thought-provoking of the events within the stories?

3. Which adventure would you like to have had with Stanley and why?

4. Where do you picture yourself in the action most? How does it make you feel to be in the middle of that particular adventure?

5. If you had a free ticket to anywhere in the world, where would you go and what would you do?

6. What kinds of issues or perspectives on life came up for you in any one story in this book?

7. What is your personal perspective on the issue you chose?

8. How would you solve any of the problems you read about?